· SKELETONS

JAN ZALASIEWICZ
AND MARK WILLIAMS

SKELETONS

the frame of life

OXFORD
UNIVERSITY PRESS

OXFORD
UNIVERSITY PRESS

Great Clarendon Street, Oxford, OX2 6DP,
United Kingdom

Oxford University Press is a department of the University of Oxford.
It furthers the University's objective of excellence in research, scholarship,
and education by publishing worldwide. Oxford is a registered trade mark of
Oxford University Press in the UK and in certain other countries

Published in the United States of America by Oxford University Press
198 Madison Avenue, New York, NY 10016, United States of America

British Library Cataloguing in Publication Data
Data available

Library of Congress Control Number: 2017953468

ISBN 978–0–19–880210–5

Printed in Great Britain by
Clays Ltd, St Ives plc

Links to third party websites are provided by Oxford in good faith and
for information only. Oxford disclaims any responsibility for the materials
contained in any third party website referenced in this work.

To Adrian Rushton
nonpareil colleague and scholar of ancient skeletons

CONTENTS

ACKNOWLEDGEMENTS

This book arose, as did the previous ones we have written, out of a moment of improvisation, which then somehow had to be carried forwards to some kind of conclusion. The idea was to take skeletons, the (mostly) hidden framework of biological life, which, in ancient form, has also provided the framework—in somewhat more exposed and visible form—of much of our professional lives, and make sense of them in the way that they have underpinned the development of complex life on Earth.

In making this story take shape, we are deeply grateful for the unceasing support and encouragement of Latha Menon and others at Oxford University Press. Latha has an unrivalled ability to see where a narrative line might be heading for the rocks, and to gently steer it into more productive waters, while Jenny Nugee and her colleagues have been cheerfully patient with our perilously close encounters with deadlines, and marvellously efficient in compiling the finished product.

As to the subject matter, the content here has been influenced by all the people who have influenced *us* as we have come to grips with fossil skeletons, old and new, in our studies and in our working lives. That is an awfully long list, but includes colleagues such as David and Derek Siveter, Richard Fortey, and the late and much missed figures of Dick Aldridge and Barrie Rickards. We dedicate this book to Adrian Rushton, who has influenced these pages both directly and indirectly; the range of fossil skeletons he has covered, expertly, has not been rivalled since the days of the great Victorian-era polymaths, and the amount of help and support he has given to others through his career has been immeasurable. We also thank the following for the images that we have used in this book and for advice on images: John Ahlgren, Peiyun Cong, Iván Cortijo, Jason Dunlop, Dennis Hansen, Tom Harvey, Soraya Marali, David Martill,

ACKNOWLEDGEMENTS

Giles Miller, Chris Nedza, Mark Purnell, Adrian Rushton, Paul Selden, David Siveter, Derek Siveter, Vincent Perrier, Ulrich Salzmann, Bernd Schöne, Ian Wilkinson, Xianguang Hou, Xiaoya Ma, and Jeremy Young.

We thank our families, too—Asih, Kasia, Milana, Mat—for the support and inspiration that they continually give—and infinite patience, too, as we steal time to write this book. And our parents too, Doreen, Les, Irena, and Feliks, who gave us the inspiration to be inquisitive.

PROLOGUE

What links the tumbleweed that has invaded the North American plains, the single outstretched canine tooth of a narwhal, the fourth finger of a pterosaur, the carapace of a beetle, and the ancient submarine mountains, reaching 5 kilometres tall, at the top of which the coral islands of tropical oceans just emerge at the ocean surface? These are all forms of skeleton, produced by living organisms in an extraordinary late flurry of evolution on a planet that was already easing into middle age.

Imagine, then, a world where skeletons had not evolved. We would not see birds flying through the air to perch on a tree branch, or a cat leaping high onto a garden wall (perhaps to try to catch one of those birds), or a crab scuttling on a beach, or a child running through a playground. We might be forgiven for thinking that a world without such familiar things would be a bizarre place indeed. But that is just how it was for most of the history of life on Earth.

The Earth's surface now holds a vast array of skeletons, from microscopic to gigantic. Some piles of skeletons, like the Great Barrier Reef of Australia, are so large that they can be seen from space, while others, visible only through powerful microscopes, show exquisite preservation of minute structures from hundreds of millions of years ago. Specimens of the most spectacular skeletons have been avidly sought by humans, even in antiquity, and their power to awe remains undimmed. Looking from another perspective, skeletons *en masse*, assembled down the years and stored within layers of rock, have been crucial in controlling some of the Earth's most important chemical cycles, and in maintaining a habitable climate on our planet.

In this book we look at skeletons from many angles, and encompass all of those mineral frameworks that have allowed life to engineer the planet we live on. We look at the skeleton innovations that occurred in deep antiquity, within tiny cells that built forms of scaffolding for themselves, and how this led to the eventual evolution of the familiar skeletons of today. We explain the very different strategies by which skeletons can be built, and we show how these frameworks can engineer an extraordinary diversity of bodies, shapes, sizes, and solutions to the problem of living.

Looking at life on Earth from the perspective of skeletons can help to answer some big questions. Why are skeletons made from certain materials? Why do some animals have their skeletons on the outside, and others on the inside? What advantages have skeletons conferred to animals and plants, and what lifestyle possibilities have they enabled? And how have they re-engineered our planet, in providing the framework for sophisticated networks of life that fashioned the evolution of Earth's oceans, land, and atmosphere.

How have skeletons, too, helped *human* life? The bones of dinosaurs and marine reptiles are not only prized as spectacular museum pieces, to give us glimpses of the Earth's ancient and dramatic past. Farmers have used them, too, in a much more prosaic and functional way, albeit in a way that would dismay a palaeontologist. And, now, as we enhance our own skeletons technologically, and build new skeletons to allow us to explore our world and other worlds, scientists can look to the structure of bone and shell that has evolved over millions of years, to provide inspiration as they construct the new material fabrics of our society.

To humans, skeletons have long been more profound symbols, too, of both death and permanence, and this resonance has undoubtedly increased their fascination. In the course of this book, we will come across some of the scientists who have discovered and studied spectacular skeletons, living and fossil, and worked to solve the many mysteries of life and death on Earth: of how that life came to flourish and develop—and how it crashed when those skeleton-building powers periodically failed.

What might be the future for skeletons as the Earth grows old? The inexorable course of our Sun's evolution as a star means that there is about another billion years for this particular experiment of Earth-bound life to run. There is time, therefore, for the story of skeletons on Earth to develop new chapters yet—and, as we write, new chapters seem to be beginning. We seem to be on the verge of repeating ancient catastrophes to the skeletal fabric of our world, and recreating some of the nightmare worlds of the past, as a gift to our children.

Yet, simultaneously, our species is producing extraordinary novelty. What might be the future of a planet where skeletons are fashioned of metal, plastic, silicon, and re-engineered tissue and bone? Perhaps our new, invented frameworks for life will help us travel towards distant stars and planets to encounter different organisms with different skeletons? After the past few tumultuous years of discovery, such odysseys, on this world and others, might now be realistically contemplated, with both foreboding and anticipation. Wherever they will take us, skeletons of one sort or another will establish the framework for future evolution—and endure thereafter to tell the story.

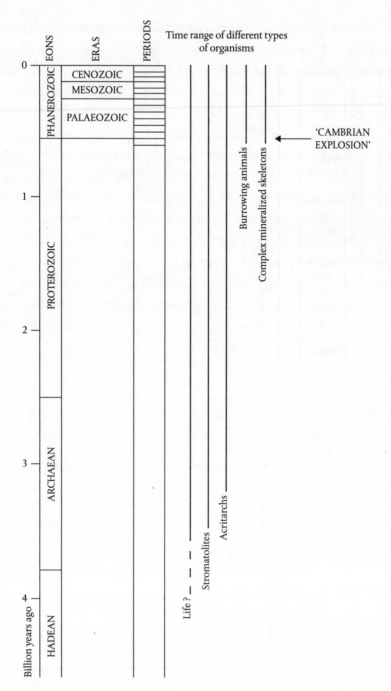

Geological timeline.

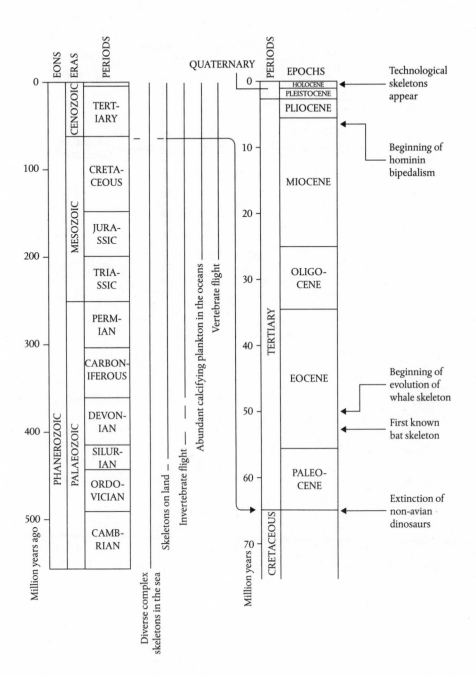

1

SKELETONS APPEAR

Charles Darwin saw a world of two halves. It perplexed him deeply. He looked at the familiar world around him where cats walked and leaped, insects and pigeons flew, fish and whales swam in the seas, and shellfish abounded on sea floors and along coastlines. As a young scientist circumnavigating the globe on the *Beagle*, he had collected fossilized shells in the Canary Islands and the bones of gigantic extinct sloths and armadillos on the coast of Argentina. Even before that, as a Cambridge undergraduate student, he had been in the field with that mighty geologist of Victorian days, Adam Sedgwick, a man who had extracted countless fossil trilobite skeletons from rock strata formed during the Cambrian Period that long predated the appearance of bony animals on land.

Yet, Darwin knew that below these Cambrian rock strata that held fossil skeletons of many different kinds, there were enormous thicknesses of yet older rocks of Precambrian times, in which there were no bones and no shells. How had this ancient world without skeletons changed—suddenly, it seemed—to a world abounding in these most durable and versatile frameworks for life?

It is a dilemma that scientists still puzzle over, although the problem is now framed differently. Perhaps, in some ways, it is even greater. Darwin could see no evidence for life of any kind in the ancient rocks of the Precambrian. For all he knew, that world might have been a dead planet. And, he had no way of measuring the time spans involved, in order to assess how long ago this all was. Because we have leant to read the atomic clocks built into some specific minerals that can be found in rocks, we know that the Earth is 4.54 billion years old, that traces of microbial life can be found in rocks at least 3.8 billion years old and perhaps even as old as 4.1 billion years—and that the skeleton-packed Cambrian Period began just 541 million years ago. Life has existed therefore for at least fourth-fifths—and perhaps nine-tenths—of our planet's duration. Yet thoroughgoing, familiar, skeletons, and the kind of life that they literally supported, occupy just the past 12% of Earth time. Skeletons may, on a planetary timescale, be a latecomer innovation, but they changed the world fundamentally. Indeed, they define the present geological eon, the Phanerozoic, of which the Cambrian is the lowest rung.

The problem becomes yet more curious in that we now know too, that for almost all of the time that life has existed on Earth, microbes could indeed produce skeletons of a kind that were either too tiny for Darwin to see or too crude for him to recognize. We know, too, from exceedingly rare fossils, that soft-bodied multicellular organisms have existed on Earth for at least 1.2 billion years, and that a distinctive array of mysterious, multicellular but seemingly skeleton-less organisms, the 'Ediacara biota', became quite widespread in the oceans about 60 million years before the start of the Cambrian Period.

Whatever the skeleton factor is, it is clearly pretty special—and when little *Cloudina* appeared on Earth about 550 million years ago, with its brand new shell, it was the start of a revolution on the planet.

The First True Skeletons

Preston Cloud would be included in any sensible catalogue of remarkable geologists. As a young man in the US Navy of the early 1930s, he became that organization's bantamweight boxing champion. Later, not able to attend university due to the Great Depression, he went to night school and did manual work at the US National Museum during the day. An enthusiasm for fossils got him a job as a preparator in the palaeontology laboratory at the museum, from which opportunity his career blossomed. A man with a wide-angle view of the Earth and its place in the cosmos, he is best known for deciphering, in memorable terms, the larger structure of Precambrian history—he was the one to coin the term 'Hadean Eon' to mark the earliest, most mysterious part of Precambrian time, for instance. So it is fitting that one of the earliest known truly skeleton-bearing organisms on Earth is named after him (Figure 1).

Cloudina (of the family *Cloudinidae*, just to add further lustre to the great man) is hugely important, but it is not at all the most spectacular of fossils. It is basically a rather irregular curved tube, up to a few millimetres wide and a few centimetres long. The tube has a distinctive structure, being a set of long cones, stacked one inside the other; the first cone is closed at its narrow end, and all the others are open.[1] And

Figure 1. *Cloudina carinata*, late Ediacaran (Precambrian), El Membrillar olistostrome, Helechosa de los Montes, Badajoz Province, Extremadura Region, Spain. Scale bar is 5 mm.

that's it: there are no structures such as holdfasts, no internal partitions, and no lid. *Cloudina* is moderately common in strata of the very latest Precambrian, from about 550 million years ago. It coexisted with the last of the enigmatic Ediacaran organisms—but they are not found together in the same rock strata, so perhaps they lived on different parts of the sea floor. *Cloudina* has a reasonable claim to represent the beginning of skeleton formation as we commonly understand it.

The tubes seem to have been constructed of calcite, the common form of calcium carbonate. Discerning the original composition of these very early skeletons, of *Cloudina* and the forms that were to shortly succeed it, is not always easy. Perhaps because of the chemical makeup of late Precambrian seas, the shell material was prone to being altered after death, on the sea floor or just following burial by sediment. The replacing material is often calcium phosphate, though in other examples silica is involved, or alternatively calcium carbonate. In the case of *Cloudina*, well-preserved examples seem to show that the original shell material was of calcite, in tiny, micrometre-sized crystals. These crystals seem to have originally been embedded in some sort of tough exterior organic material, for the somewhat corrugated appearance of the tubes suggests that they were moderately flexible in life, though they became brittle after death and recrystallization.

What kind of biochemical trick did *Cloudina* use to make its skeleton? The basic mechanism of all such *biomineralization* in all skeleton-builders hinges on modifying conditions inside the tissues to encourage chemicals that are generally dissolved in fluid in and around the organism to crystallize out as hard structures. The sea, in general, is so saturated with ions of calcium (Ca^{2+}) and carbonate (CO_3^{2-}) that calcium carbonate ($CaCO_3$) can easily crystallize out of it—with no help from biology at all. Organisms can then encourage this process by tweaking the chemical conditions around them, by concentrating the ingredients. Or they can change factors that affect crystallization, of which one of the most important is the acidity, which is measured on the pH scale, a measure of the concentration of hydrogen ions (H^+). The lower the pH—that is the more hydrogen ions, as it's an inverse relation—the more acid are the conditions, discouraging crystallization. Raise the pH in the tissues, though, to give more alkaline conditions, and crystallization is encouraged. *Cloudina* had clearly evolved an effective biomineralizing mechanism. It had a makeshift skeleton, compared with some of the marvels that we will consider later. Nevertheless, a thoroughgoing skeleton it was. Why go to the effort to build it?

Attack of the Soft Animals

Cloudina appeared at about the same time as another major innovation in the Earth's biosphere. About 550 million years ago, strata also begin to show the first signs of substantial burrowing, indicating the evolution of animals sufficiently muscular and active to push through the sediments of the sea floor. These animals punctured or churned the surface of the sediment in a process that geologists call bioturbation, to produce a mottled, disturbed texture in the rocks. There are occasionally a few simpler traces in older Precambrian rocks, but nothing of this scale and complexity. From the beginning of the Cambrian, though, this kind of bioturbated rock texture becomes common. To push through the sea floor sediments in this way, a special kind of skeleton was needed, one with a recognizable head and tail end, of a *bilaterian* animal.

More primitive animals, such as jellyfish, possess a simple, soft hydrostatic skeleton, a fluid-filled interior, surrounded by muscular layers under the control of a nervous system. As the fluid here—seawater—is incompressible, contraction of muscles that may be arranged either around the animal, or along its length, will change its shape. That is only a beginning of such muscularity. A jellyfish can move in this way through seawater—but it cannot burrow through sediment.[2]

With a more advanced body plan—one with a flexible but tough external cuticle—the animal can do just that, as any glance at an earthworm in garden soil will show (Darwin, who made a detailed study of earthworms, had as good an insight into these processes as anyone). That kind of advance was only possible with the evolution of a more sophisticated and robust array of structural tissues, in the body plan of a bilaterian animal. Bilaterians include most major animal groups, and the appearance of bioturbated strata 550 million years ago is evidence of when this major step within evolution took place.

This truly was a major step, perhaps taken in response to the invention of predation. For though jellyfish and sea anemones may possess muscle tissues, and nerves, and a simple gut, at the embryonic stage they only possess two germ layers of tissue and this limits the range of structures they can produce. These animals are called diploblasts, and some of them, like corals, nevertheless went on to build the greatest skeletal structures that have ever functioned. Most animals, from worms to humans, are triploblasts, in that the embryo possesses three layers of germ cells. This evolutionary step was of crucial importance for skeletons, because it allowed triploblasts to make a much wider range of structures, including organs. But for our story it enabled the development of the internal bony skeleton of humans, which is derived from the mesoderm germ layer, and the mineralized exoskeleton of a trilobite, which is derived from the outer layer, the ectoderm. And crucially, it produces the unmineralized but nevertheless tough outer cuticle of a worm, of the kind of organism that first began tunnelling into the Precambrian sea floor.

This revolution marks our eon, for it heralded the transformation of the biosphere shortly to follow, and, as one result, made hard skeletons a crucial part of the biological toolkit for survival. The body of bilaterians was a major factor—perhaps *the* major factor—in kick-starting this biospheric revolution, and as regards our narrative it provided the strongest of selective pressures to encourage development of the solid skeletons that soon followed. It was an innovation fit to mark an eon.

The formal boundary for the beginning of the Phanerozoic Eon—the one we still live in—was chosen for a particularly distinctive kind of early burrow, which represented a kind of three-dimensional corkscrewing movement, and named *Treptichnus pedum*, first seen to appear within sea floor sediments that now comprise strata in Newfoundland, Canada. These burrows are similar to burrows made today by a priapulid worm, as it corkscrews through sea floor sediment in search of prey. This new trick of muscular movement was thought to represent the most consistently traceable level for the eon's beginning and, although it is has proved a little problematic in practice,[3] that is the boundary which, for now, remains in place.

Any animal then possessing the ability to crawl, slither, or burrow could exploit the food reserves in organic-rich mud, simply by eating that mud, while microbial mats could be treated as food too. At that delicate point, the stage was set for further developments. That muscularity, for many of those early forms, began to be directed towards the hunting and eating of other organisms, particularly if they were just a little smaller and a little less mobile. The evolutionary arms race, for so long slow and quiescent—as far as we can decipher—had begun in earnest. *Cloudina* was the first defensive riposte that we know of in this arms race.

The evidence is clear in the fossils themselves. In well-preserved assemblages of these skeletons, up to one-fifth of *Cloudina* tubes may show evidence of attack, in the form of neat circular puncture holes in the shells.[4] Some of the holes are incomplete—they did not penetrate all the shell, and others were unsuccessful, for the *Cloudina* individual lived to be punctured again later in its life—sometimes several times. The shell therefore undoubtedly formed a protective armour that increased the chances of survival to the skeleton-maker.

Even at this early stage in the arms race, though, the ecology was not simple. There seems to have been just one predator in the examples studied, for there appears to be just one type of puncture mark. But there was another potential target. *Cloudina* assemblages may be accompanied in places by another, similar tubular shell, *Sinotubulites* (which differs by having a two-layered tube that is open at both ends[5]). Where these coexist and there is evidence of predation, it is *Cloudina* that is the victim, while *Sinotubulites* shells have—in the specimens examined to date, at least—shown no signs of attack. The early predator seems to have been selective.

What kept *Sinotubulites* safe? Perhaps it secreted toxins to keep the unknown predator at bay, or had additional protective armour made of organic material, that was not preserved. In the arms race made possible by mobility and armour, complexity was there right from the start—and it was only to increase, with the extraordinary outburst of skeleton construction that was, very shortly, to follow.

An Eruption of Skeletons

'Darwin's dilemma' marked the sudden change from ancient barren strata, seemingly devoid of signs of life, to more recent strata abounding in the remains of shells and carapaces, of which *Cloudina* and *Sinotubulites* were just the advance guard. Charles Darwin saw this absence as a major problem in his attempts to explain how life evolved on Earth—and would have been even more puzzled had he been aware that the Earth nurtured skeleton-bearing life for a mere one-eighth of its history. He would have expected to see the fossilized remains of more primitive ancestors, extending back into the mists of time. To him complex, skeletonized, fossilizable life seemed to come in abruptly at the 'explosion' of fossils at the beginning of the Cambrian.

However, after more than a century of painstaking study of the strata, we can now ask more detailed questions about this Cambrian explosion of life. Quite how sudden was it, and how closely tied to the acquisition of skeletons? Was the explosion essentially one of easily fossilizable skeletons that followed a long, cryptic history of complex, soft, multi-cellular organisms? Did skeleton-making arise just once, or did it appear independently in different evolutionary lines of organisms? Was the ability to make skeletons triggered by the development of sufficiently sophisticated biology, or was it caused by some kind of environmental changes, to make skeleton-making easier? Or by both? There is a plethora of questions here, based around the complex relation of skeletons to biology and to planetary conditions. We can unpick them—some of them, at least—one by one.

The fossil record is much better known than it was in Darwin's time (Darwin himself was rather deprecating about the use of fossils in providing evidence for what he called 'descent with modification' and what we now know as biological evolution; in *The Origin of Species*, he placed more weight on evidence from animal breeding). Crucially, the fossil record has now been numerically calibrated—that is, it has been dated in numbers of millions of years, using radiometric ages. Amid the skeleton-bearing strata, there are, here and there, layers of volcanic ash. These commonly contain crystals of minerals such as zircon (zirconium silicate) and monazite (a phosphate of rare earth elements) which, when they grew, included significant amounts of the radioactive element uranium. Radioactive decay of the uranium to lead within the crystal is the basis of a highly effective atomic clock, which in good circumstances can establish the age of a stratum to the nearest million years or so—even at a distance of half a billion years.

Cloudina, Sinotubulites, and others—those first true shell-formers—followed shortly after the very first burrows, and were themselves followed soon after, in geological terms, by the distinctive three-dimensional fossilized burrows of *Treptichnus pedum*, chosen to mark the official beginning of the Cambrian Period and simultaneously the Phanerozoic Eon (and simultaneously too, of the intermediate category of time division, the Palaeozoic Era that ended catastrophically 252 million years ago). Strata laid down over the next few million years of the Cambrian Period include the next stage in skeleton formation. These are the 'small shelly fossils'—a variety of minute button-like, or tube-like, or shell-like fossils, which typically range from fractions of a millimetre to a few millimetres in size. These are disarticulated remains, like a broken-up suite of chainmail armour. And as a result, it is often difficult to know what animal they belonged to. Many of these animals remain puzzling, but some have, by careful study, been related to different types of organism. The Cambrian explosion was gathering pace.

Then, a little more than 520 million years ago, there appeared the signature fossil of the Cambrian, the trilobite (Figure 2). To many people,

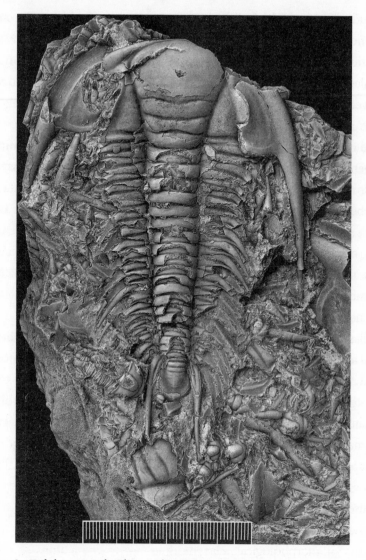

Figure 2. Trilobite, *Paradoxides paradoxissimus* preserved in limestone, Varnhem, 12 km E of Skara, Vastergotland, Sweden. Scale bar is 5 cm.

these charismatic fossils symbolize palaeontology perhaps even more than do the dinosaurs and the mammoths. With their large, lustrous, calcified external skeleton, subdivided into head, body, and tail, commonly armed with prominent spinose projections, they are unlike any animal living, and have attracted attention long before they began to be studied palaeontologically. They were prized as decorative objects by the ancient Chinese, Greeks, and Romans—and also by the Ute Native Americans, who extracted superb specimens of the trilobite now known to paleontologists as *Elrathia kingi* from the rock layers of Utah, and who believed that amulets made of these 'little water bugs' guarded against illness and bullets. This allure to the human sense of the aesthetic (and, in the case of the Ute people, the spiritual) is a reflection of the complex, segmented trilobite armour, which has a skeletal sophistication far beyond the simple tubes of *Cloudina*. Trilobites can be found even in rock strata that have been strongly altered by heat and pressure underground—such as the Welsh slates that Adam Sedgwick studied. Their skeletons are emblematic of the change that came in with the advent of Cambrian time, and they are often the only fossils to be commonly found when palaeontologists hammer their way through early Cambrian rock successions.

Other types of skeleton appeared around the same time. As well as the 'small shelly fossils', which include some enigmatic, if not downright weird animals, there were the first echinoderms, brachiopods (sometimes called 'lamp shells'), gastropods, and archaeocyathids—extinct sponge-like organisms whose robust tube-like skeletons could pile up so thickly that they formed the first true reefs of the ocean. It was quite a menagerie. The main burst of skeleton formation in the Cambrian explosion took perhaps as long as 30 million years—roughly half the time that separates us from the dinosaurs. Preston Cloud called it the 'Cambrian eruption' because of its unfolding over such a long duration—and also because, as he wryly noted, 'it didn't make any noise'. Nevertheless, following billions of years of morphological near stasis in Precambrian time, it was an extraordinary, and exceedingly rapid, burst of skeletal innovation, which effectively determined the pattern of life and of evolution up to the present day. Virtually all of the major body plans of living organisms—including that of the vertebrates—appeared in this radiation, and this has determined the development of skeleton types to this day.

Rapid though this evolution was, it was not *unreasonably* rapid.[6] Darwin's famous misgivings about this event, about it not being explicable by

11

normal evolutionary processes, can be allayed, although that has not stopped the generation of adventurous ideas. It is known that evolution is not a one-speed process, whether measured in terms of observable change in size and shape between older and younger fossilized organisms, or as change inferred from genetic patterns in modern organisms. For instance, small islands do not usually have enough resources to support large mammals, and so any large mammal populations that do find themselves on small islands—trapped there by some change in sea level, perhaps—commonly undergo 'island dwarfing', markedly and rapidly shrinking in size. Fossils show that this can happen over only thousands, rather than millions, of years (the last known mammoths, on Wrangel Island north of Siberia, were not much bigger than a pony for this reason, before they finally died out some 4000 years ago[7]). The appearance of major biological innovations, too, leads to rapid evolution, as the new features are quickly honed—and as the other species affected by them adapt in response. The Cambrian explosion, or eruption, was a time of extraordinary and major biological innovation, and rates of evolution were heightened to match that.

Harder and Softer Skeletons

There is something of a catch-22 here. If all of our direct evidence of past life comes from fossils, then it is restricted to those organisms that fossilize, and hence to those organisms with skeletons of one sort or another. These skeletons may possess a complete body armour as in some organisms, or just one or two skeletal parts in an otherwise soft-bodied organism, like the hard resistant teeth of mostly soft-bodied sharks. No matter—each will leave easily traceable clues to their former owners in the strata. But what about organisms that are wholly soft-bodied? How much of the evolutionary story of the Cambrian is hidden from us because of our understandable focus on skeletons?

Luckily, we now have some measure of this, because Cambrian rock successions are—for reasons that have been much debated but are still a little obscure—rich in those prized fossil localities where both soft and hard tissues have been petrified, to give a more complete picture of the biology of those times: 'windows on life', as they have sometimes been called, or, more technically, conservation 'Lagerstätten' (the word derives from the German for 'storage place'. The word is usually capitalized, as are all nouns in German, and the singular form is 'Lagerstätte').

Two such Lagerstätten are of prime importance to this story. One is the Burgess Shale, in British Columbia in Canada, and the other is the equally astonishing, more recently discovered Chengjiang Biota of Yunnan Province, China. The Burgess Shale was discovered by the celebrated geologist Charles Walcott. This particular shale lies in high, rugged terrain, on the steep flank of Mount Stephen. Walcott first heard in 1886 that magnificent trilobite specimens had come from thereabouts, but it was only in 1907 that he managed to explore the mountain for himself; the visit was brief, but he found some fine trilobites, encouraging him to go back and look for more. And in 1909 he first found fossils of a kind never seen before—including sponges and baroque crustacean-like animals, their legs and antennae finely preserved. Knowing this was a major find, he went back each summer, complete with wife and children, who made the steep trek up the mountainside with him on horseback. Over several summers this improvised family business opened up a respectably sized quarry on the mountain side. Using hammers, chisels, large crowbars—and sometimes dynamite—large slabs of shale were levered out, slid down the slope and split open for their treasure trove of magnificent fossils, of which some 65 000 specimens eventually made their way back to the Smithsonian Institution in Washington, where they still reside, neatly arranged in drawers with their original labels.

The Burgess Shale represents the middle part of the Cambrian Period, about 500 million years ago, and some time after the main detonation of the Cambrian explosion. The Chengjiang deposits are even older, dating back to just after the appearance of trilobites on the Earth. These were discovered, serendipitously,[8] by a young geologist, Xianguang Hou, who at that time was working at the Nanjing Institute of Geology and Palaeontology. The fossils, with their exquisite preservation of both soft and hard tissues, are preserved in vivid orange and brown on the pale yellow rock, contrasting nicely with the silvery-fossil-on-black-shale preservation common in the Burgess Shale. In both cases, the high-fidelity preservation is at least partly due to both of these rock successions representing mass burial events, when portions of Cambrian sea floors where the animals lived were suddenly overwhelmed by catastrophic slurries of dense mud.[9] Thus entombed, the carcasses were preserved from both scavenging and decay—until brought into the light again by the questing palaeontologists of our era. Given this much more representative sampling of the biology of the Cambrian sea floor—how common, or how rare, were animals with hard skeletons compared to soft-bodied ones? Were they still just rarities among a throng of soft-bodied animals that we happen to find commonly as fossils because of their hard shells—or were they really abundant at those times?

Based on systematic studies of tens of thousands of specimens from both of these Lagerstätten, the pattern is clear. Even just after the main phase of the Cambrian explosion, skeletons were very much the order of the day on Cambrian sea floors. Three main sorts of skeleton could be recognized, and between them they accounted for most of the fossils present. First, there were animals like the trilobites and brachiopods, with hard biomineralized external skeletons. These are the kind of 'hard-shelled' animals that are commonly found as fossils in 'normal' strata, and not just in rare Lagerstätten.

Then, there are animals, including many arthropods, which have skeletons every bit as biologically real as those of their hard-shelled kin that act as support for muscles in exactly the same way—but which differ simply in not being biomineralized, in lacking that extra hardening by (usually) calcium carbonate or calcium phosphate. This kind of non-biomineralized skeleton is rarely found in normal strata, and it needs the enhanced fossil-preserving qualities of a Lagerstätte to illuminate the former abundance and diversity of the animals that were so lightly armoured. Today, such a delicate, non-biomineralized skeleton is typical of many shrimps and of animals like krill, which are hugely abundant, but which have a very sparse fossil record. Even in animals with mineralized skeletons such as the trilobites, there were portions, such as the limbs, which lacked such armouring and so are typically only seen in Lagerstätte specimens.

Then, there are kinds of 'soft' skeletons that we have noted above, where bilaterian animals such as worms possess a tough outer cuticle that is the framework for the soft tissues within. These also are rarely preserved as fossils in normal circumstances. (They are also rarely considered as 'skeletons' in the common understanding of the word.)

Arthropods, with their external or exoskeletons, are the dominant fossils in both the Chengjiang and Burgess shales,[10] in terms of both numbers of individuals and numbers of species recognized (Figure 3). Most of them possess non-biomineralized skeletons and so are only 'visible' in these Lagerstätten. Other common skeletal fossils were the

Figure 3. Fossil arthropod *Chuandianella ovata*, lower Cambrian, Chengjiang Biota, Yunnan Province, China. Horizontal field of view is 3 cm. The head-end of the animal is to the left, with the segmented trunk to the right.

Figure 4. Fossil worms, *Maotianshania cyclindrica*, lower Cambrian, Chengjiang Biota, Yunnan Province, China. Horizontal field of view is 2.3 cm.

brachiopods, sponges, hyolithids (enigmatic animals with a conical shell, which have recently been allied with the brachiopods[11]) and molluscs. Worms with their cuticular skeletons (Figure 4) made a strong showing, particularly the priapulid worms (so named for their fanciful resemblance to a human penis) that we have already mentioned. Indeed, in terms of individuals, the priapulids formed the dominant element in collections made at a few levels within the Chengjiang deposits. Other animals included lobopods, which also have a flexible, sclerotized external skeleton rather like that of worms, sometimes with mineralized sclerites that make small shelly fossils (Figure 5). Truly soft-bodied animals like the cnidarians (sea anemones) and ctenophores (comb jellies) are rarely found in these Lagerstätten (Figure 6). The chordates, from which the vertebrates were to arise, are great rarities in these strata too.

Figure 5. Fossil lobopod, *Microdictyon sinicum*, lower Cambrian, Chengjiang Biota, Yunnan Province, China. The narrow coiled part of the body to the left is assumed to be the head end. Horizontal field of view is 2 cm.

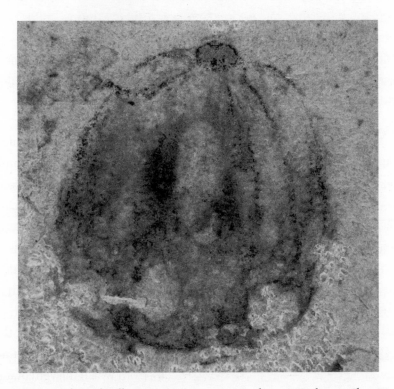

Figure 6. Fossil comb jelly, *Maotianoascus octonarius*, lower Cambrian, Chengjiang Biota, Yunnan Province, China. Horizontal field of view is 0.78 cm.

The skeleton factor, therefore, clearly played a large part from very near the beginning of complex multicellular life. Both the 'normal' fossil record and exceptional localities like those of the Burgess and Chengjiang mudrocks have given us powerful insights into how skeleton-bearing animals were an integral part, and not a bit player, amid the pattern of life in Cambrian times.

Primordial Skeletons of Lilliput?

Michael Welland's book *Sand*[12] shows beautifully how ordinary material can be extraordinary when viewed through an appropriate prism. Among the most striking aspects of sand are not the grains themselves, although the physics of sand grain movement is both beautiful and complex. It is what can be found between them, among grains on a beach or in a shallow sea. Making a living in the spaces between the sand grains are a multitude of tiny animals, a fraction—often a *very* small fraction—of a millimetre long. There is a whole gallery of them, most with entirely unfamiliar names: gastrotrichs, kinorhynchs, loriciferans, gnathostomulids, and more.

The unfamiliarity extends to science. The loriciferans, for example, were recognized as a distinct category only in 1983, and yet they represent a separate animal phylum. Since that time, 36 species have been described, at least a hundred new forms are present, not yet described, in museum collections, and perhaps as many as a thousand further species are out there in the world, as yet unrecognized. The loriciferans have their own chitinous outer skeleton, looking a little like a microscopic ice-cream cone, which they anchor so firmly to sediment grains that they can't be shaken off (a major reason for their late discovery), and from this cone extends a mass of feeding arms.

The kinorhynchs are relatives of the loriciferans and equally tiny, though have been given the more ferocious-sounding nickname of mud dragons, as they can live between mud as well as sand particles.

Kinorhynchs have a microscopic segmented body, with a tough cuticle exterior serving as exoskeleton, from which extend spines that are used to help the animal move around. The gnathostomulids by contrast are soft-bodied, worm-like in appearance—though they have minute but tough 'teeth' made of chitin.

These, and other Lilliputian animals, make up quite a menagerie, and are abundant on Earth (similar assemblages are found in soils), if mostly unnoticed by we Brobdignadian humans. Collectively they form the meiofauna.[13] The crucial thing about them is that although they are not much larger than many single-celled protozoans (and are indeed, smaller than some), they are complete, if highly miniaturized animals, with tissue layers and organs—and many have thoroughgoing, elaborate skeletons too, albeit ones that are too tiny and too weakly mineralized to be capable of being easily fossilized. This has led to the meiofauna being suggested as one answer to the key question of the Cambrian explosion: how complex, skeleton-bearing multicellular organisms seem to have appeared over a relatively short time, geologically speaking, already fully formed. 'Darwin's dilemma' had been compounded because of evidence arising out of a 'molecular clock', where analysis of the molecular structure of the genes of living organisms has suggested that the major branches of animals began to split off from each other some 800 million years ago.

Perhaps, it has been said, the visible, palaeontological record of evolution of large, multicellular animals with complex, hard mineralized skeletons like the trilobites was preceded by tens or even hundreds of millions of years of evolution of such meiofaunal animals, so tiny and with such delicate skeletons that they would be almost unfossilizable under normal conditions, and so would be hidden from us?[14]

It is a lovely, enticing idea, of evolution in a Lilliput animal world of the late Precambrian. And one aspect of some of the meiofaunal animals has been seen as offering support for such an early, hidden phase of evolution of tiny animals with diaphanous skeletons: the ability of some of them to survive under conditions with very little oxygen—conditions that would have been more widespread in Precambrian seas. This has led to much searching of late Precambrian strata for some 'meiofauana Lagerstätte' that might yield exceptionally preserved examples of such minuscule, delicate fossils—and hence give proof to this tempting theory.

Alas, the Precambrian has so far stubbornly refused to yield any such fossils. It has yielded other fossils that one would have thought yet more delicate and unfossilizable, such as fossilized embryos of some still unknown organism in 570-million-year-old rocks in China. And, conversely, the palaeontologists Tom Harvey and Nick Butterfield have recently found a beautifully preserved fossil of an undoubted loriciferan, with its delicate skeleton-cone containing a tangled mass of its (soft) arms too, in Cambrian strata from Canada.[15] Hence, such miniaturized animals can indeed be fossilized, albeit rarely.

It seems likely, then, that the Cambrian explosion (or eruption) is exactly what the name conveys—a phenomenon where the major groups of animals arose, and developed their skeletons (whether full size or miniature), essentially in Cambrian times.

It is time to look at that variety of skeletons more closely.

2

A SHELL ON THE OUTSIDE

In the summer of 1945, in the part of the New Mexico desert called the 'Jornado del Muerto'– which translates as the 'journey of the dead man'—the world's first atomic bomb exploded. Some time later, giant human-eating ants climbed out of the desert to terrorize the inhabitants of Los Angeles, building a colossal underground nest in the city's drains and sewers. This, at least, is what happened in the 1954 sci-fi classic Hollywood movie *Them*. The idea of giant arthropods, mutating after scientists have tinkered with their biology, did not stop with the ants. In 1955, a 30-metre-high tarantula rampaged out of the Arizona desert in the film titled, predictably enough, *Tarantula!* It had been re-engineered by scientists whose inventiveness also stretched to giant rabbits and enormous guinea pigs. After consuming a few cows and people, this spectacular arachnid was finally itself consumed, by fire, napalmed by the US Air Force. The giant arthropod theme has been reprised: the *Starship Troopers* battled an endless supply of oversized insects from outer space, while just to turn the tables, in the film *Aliens* Sigourney Weaver dons an artificial exoskeleton to do final battle with the monstrous queen Alien.

Is the possibility of gigantic man-eating bugs a realistic proposition? Looking back at the truly huge organisms that once roamed—and indeed still roam—our planet, did any of them have their skeletons on the outside? The answer, luckily, is 'no'. Because, 30 metres of angry *Tarantula* with its giant pincer-like claws would likely have made short work of any animal with an internal skeleton, including elephants, mammoths, and

Tyrannosaurus rex. Animals with their skeletons on the outside—those with exoskeletons—have a limited size range to which they can grow, on this planet at least. How big is big, then, for different types of skeleton?

How Big is Big?

'Bigness' in skeletons is relative. For a tiny, unicellular foraminifer— a relative of an amoeba with a shell—to have a skeleton over 1 centimetre in diameter is enormous. Yet, just such giant foraminifer skeletons— called *Nummulites*—packed together in their billions, form the limestones of which the great pyramids at Giza in Egypt were constructed. We humans, though, looking from the perspective of our own vertebrate skeletons, tend to be impressed by yet bigger vertebrates. Blue whales, hence, are big, the biggest being over 30 metres long (Figure 7). Some dinosaurs were even bigger, if length is the main criterion: the South

Figure 7. The skeleton of the blue whale exhibited in the main gallery of the Natural History Museum in London.

American Cretaceous dinosaur *Argentinosaurus* may have been nearly 40 metres long. Such 'big' animals grow their skeletons on the inside.

How big can an animal become, though, if its skeleton is on the outside? The empirical answer seems to be about 3 metres. This is with a whole-body exoskeleton. If the external shell is reduced or lost through evolution, then the constraints are relaxed, and the animal can grow larger. The 'colossal squid' (that really is the animal's name) is a gigantic mollusc known to grow to nearly 14 metres long. Its ancestors had an external skeleton, and the colossal squid still preserves a skeleton, though this has become internal: called a *gladius*, it resembles the shape of a Roman soldier's sword. The colossal squid is truly big, then, being somewhat larger than the biggest great white shark which is just over 6 metres long—though the squid's long tentacles make up most of its length, the body itself being 2 to 3 metres long.

Why is it that animals with their skeletons on the outside are generally—from our perspective—quite small?

The answer is a matter of engineering. Skeletons support a body's tissues, including heavy structures such as the major organs, and make points of attachment for the muscles. As an animal grows, the skeleton grows with it, thickening its supportive structures. It is here that two clear problems arise for an animal with its skeleton on the outside: how does it continue to grow within its skeleton; and how does it continue to support its body using, essentially, hollow tubes?

For animals with open-ended external skeletons such as molluscs, which add continuously to their shells as they grow bigger, the first problem is easily solved. A gastropod like the humble garden snail simply adds coils to its shell, and a bivalve such as a scallop adds new shell at its edge. Even then, it is notable that the largest living bivalve, the giant clam, is typically just over 1 metre long, whilst the largest of all ancient bivalves, the Cretaceous *Inoceramus steenstrupi*, may have reached about twice that size.

But in an animal like an arthropod, the tissues are entirely enclosed within its cuticle. Think of a spider or a lobster. Simply adding to the edges, or growing an extra coil, is no answer, and is indeed impossible. This poses a major problem for arthropods, particularly those on land, because while they are growing the animal must moult its skeleton to form a newer, larger one. Arthropods, then, are vulnerable to attack, and that might be one of the reasons for their overall small size. All the authorities in Los Angeles terrorized by 'Them' needed to do was to simply sit back and wait for the ants to moult, at which point they could have been dispatched with ease.

There is a more significant problem for those giant ants too, because their legs would literally crumple beneath them as they got bigger. Their skeleton is a thin, hollow structure, essentially an elongated long tube suspended on a series of smaller tubular legs (six in the case of ants). And therein lies the problem. As our radioactive ants grew bigger, doubling their length, their weight went up eightfold. Their tubular, hollow legs could not keep pace with this increasing weight, and the only answer—thickening the tubes—would simply produce a skeleton so heavy that the ants' muscles would fail. Poor ants! Rather than Los Angeles being a city of terrorized citizens, fleeing in all directions, it would have been a somewhat grisly but quite harmless sight: a mass of half-moulted and rotted carcasses of collapsed giant arthropods. That 30-metres-high tarantula, of course, would have been proportionately in a yet more distressed state.

Perhaps the filmmakers of *Them!* and *Tarantula* would have had better luck with a gigantic, marauding millipede, though this creature does not quite conjure up the same horror. Nevertheless, the largest terrestrial arthropod that ever lived was *Arthropleura*, a millipede that regularly grew beyond 2 metres long, but which, thankfully for us, lived 300 million years ago in the tropical forests of the Carboniferous Period. *Arthropleura* was once thought to be a ferocious predator, but more likely it was a plant eater, with rather modest mouthparts. How then, did *Arthropleura* get so big? There are probably several reasons. Firstly, given its size, *Arthropleura* may have had few natural predators in its Carboniferous homelands. After all, arthropods had a head start in the process of colonizing the land, the ancestors of *Arthropleura* being in the vanguard of this colonization during the late Silurian Period, some 100 million years earlier. They were therefore already very well adapted for the Carboniferous landscape. And, the Carboniferous atmosphere may have been much richer in oxygen too, perhaps to as much as 35%, making respiration easier for a large land-based arthropod. Though details of the respiration system in *Arthropleura* are unknown, by analogy with living millipedes, it likely breathed through two pairs of tiny surface valves or spiracles in the exoskeleton of each body segment, and these were attached to a tracheal system—itself a series of infolds of the exoskeleton that delivered oxygen to the animal's tissues. In *Arthropleura*, the tracheal system was probably attached to a long tubular heart that extended through much of the body. For terrestrial arthropods, this tracheal system, which is not able to deliver oxygen to the animal's tissues as effectively as can the lungs or gills of vertebrate animals, is another factor limiting overall body size.

But *Arthropleura* was not the largest arthropod of all time. That crown is taken by two aquatic animals: one from the Devonian Period, a giant 'sea scorpion' called *Jaekelopterus*, and the other being the living Japanese Spider Crab. *Jaekelopterus* was a fearsome creature that lived in the Late Devonian Period about 380 million years ago. Despite being a 'sea scorpion', it actually lived in freshwater lakes of Europe and North America. With its 2.5-metre-long body and nearly half-metre-long chelicerae (claws), it must have been an effective predator that likely hunted fish and other arthropods. Also looking like a monster from one of those 1950s American horror movies, but rather less ferocious in reality, is the Japanese Spider Crab, the largest of all arthropods if its appendages are included, reaching an impressive 5.5 metres in length from claw to claw. In reaching this size the crab has avoided the problem of hollow tubes and increasing weight, in that its body remains small (only about 40 centimetres long) whilst its legs grow extraordinarily gangly.

And this is why, in that epic deep-sea battle between the sperm whale and the giant squid, it will always be the whale that outclasses the squid in 'tentacle to jaw' combat. The spider crab and colossal squid may be big, if the size of their limbs is also included, though their bodies must always remain quite small. Possessing an exoskeleton precludes growing a large body—at least from the point of view of a terrestrial vertebrate. But the huge number of animals that have adopted this style of skeleton, or have secondarily adopted it in vertebrates such as the porcupine, hedgehog, and the dinosaur *Anklyosaurus*, indicate that exoskeletons are nonetheless really useful, not just as support structures, but also as armour-plating.

Armour-Plated Animals

For most of Earth history, organisms did not possess armour-plating. Its invention during the late Precambrian and early Cambrian, between about 550 and 520 million years ago, may have led to an evolutionary arms race between predator and predated, rather in the way that the discovery of metal—and artificial armour-plating—in the Bronze Age produced an arms race between different human cultures that still goes on today.

Despite the limits to size, many groups of animals make their skeletons on the outside, not just because this is a good way to support and articulate muscles, but because this armour-plating offers considerable security against attack. Many of these animals with exoskeletons are as familiar to us as snails, oysters, and the *Nautilus*, though some molluscs buck this trend—squid having a remnant shell on the inside, but no shell on the outside. Corals, brachiopods, and less commonly known groups such as pterobranchs, bryozoans, and phoronids also have their skeletons on the outside. But the most successful group of animals to have evolved exoskeletons are the Ecdysozoa, those which periodically shed their 'skins' by moulting (a process technically termed ecdysis, hence the name), and which include ants, velvet worms, priapulid worms, beetles, spiders, and crabs. To this group of 'skin-shedding' animals belong groups such as the polychaete worms, well known to fishermen, who collect lugworms by digging beneath their telltale 'worm-like' patterns on a sandy beach at falling tide. These were the patterns you may have squashed with your feet as a child.

Why then, given that building an exoskeleton limits your overall size, have so many different kinds of animal developed them? That story begins in the late Precambrian, about 550 million years ago, with some of the probable ancestors of later animals. Given that the development of some of the earliest skeletons was possibly a response to predation, with tiny holes drilled in the shells of *Cloudina*, one should revisit the Cambrian world to seek evidence of animals making a wider range of exoskeletons.

Humpty Dumpty World

As different types of animals began to make external skeletons during the early Cambrian Period, they experimented with many shapes and forms. They typically constructed these exoskeletons piecemeal, so we find them mainly as disarticulated remains, broken up by the action of sea bottom currents and decay, to be preserved in rocks as 'small shelly fossils', usually abbreviated by palaeontologists to 'SSFs'. Reassembling these broken-up remains to discover the animal's true identity needs time, patience, and much familiarity with these fossils. It is rather like all of the king's horses and all the king's men putting Humpty Dumpty back together again—though, as most of the pieces are usually missing, there is little clue to what Humpty originally looked like.

The name 'small shelly fossils' was first coined by the palaeontologists Crosbie Matthews and Vladimir Missarzhevsky in 1975. The name is a bit of a 'catch-all', in that it refers to the skeletal remains of diverse animals including brachiopods, the ancient relatives of sea urchins and velvet worms, possible molluscs, and some animals that are so bizarre that they currently defy classification. Nevertheless, it has proved popular with scientists who, understanding its limitations, use the term to help their discussions about the origins of different skeletons during latest

Precambrian and early Cambrian times. SSFs are one of nature's first extensive experiments with making many different types of complex biomineralized skeletons. While some SSFs represent the whole skeleton of a tiny presumed animal, as with *Cloudina*, most are disassembled and hence need putting back together.

One of the great pioneers of studying the disarticulated remains of these early animal skeletons was Englishman Edgar Sterling Cobbold, born in 1851, the eldest son of a surgeon. Cobbold worked as a civil engineer—he was in part responsible for engineering the impressive late Victorian dams of the Elan Valley in central Wales that supply the English city of Birmingham with much of its water. But it was really geology that he devoted himself to. In 1886, he moved to the small Shropshire town of Church Stretton, near the border with Wales. With its nearby rolling hills encompassing strata of Precambrian, Cambrian, Ordovician, and Silurian age, Church Stretton was a veritable paradise for Cobbold. Or rather, he made it so, becoming adept, over decades of work, at patiently gleaning fossils even from unpromising-looking rock layers, and then patiently extricating them from the tough rock matrix, with the help of a mounted needle and magnifying glass. He was one of those tireless, dedicated field geologists who took the broad geological visions of the great geologists of Victorian times and hammered them out into precisely detailed patterns of the history of ancient life.

Cobbold—who continued to collect fossils to within a week of his death—found, amongst many other fossils, small shelly fossils in Cambrian rocks near the hamlet of Comley. Cobbold recorded his fossil discoveries in many scientific publications. Perhaps his most significant work, published in 1921, is *The Cambrian Horizons of Comley (Shropshire) and their Brachiopoda, Pteropoda, Gasteropoda, etc.* The title is dry, but the study was meticulous and influential. Amongst his many descriptions in that work is that of the skeleton of an animal he called *Lapworthella*, after one of those grand figures of Victorian times, Charles Lapworth, Professor at Birmingham University and founder of the Ordovician System. *Lapworthella* is included among the small shelly fossils, being recorded worldwide in Cambrian rocks from the Antarctic to Mongolia. It looks like the inverted conical cloth hat worn by ancient Persian soldiers, but it is only a few millimetres long, and is made of calcium phosphate. It remains enigmatic too, because despite nearly 100 years of study, no one quite knows what kind of animal lay under the hat-like skeleton.

Some of the small shelly fossils have been assembled into the whole animal, though that has not always made us much the wiser about what they were. The grandiose-sounding chancelloriids of early Cambrian times are one such animal, with millimetre-sized sclerites (armoured plates) that were probably made of calcium carbonate (Figure 8). These sclerites—some blade-like, others star-shaped—covered the animal's body, often forming a densely packed 'crown of thorns' towards the presumed apex of the animal. *Nidelric* is one of these animals (Figure 9).

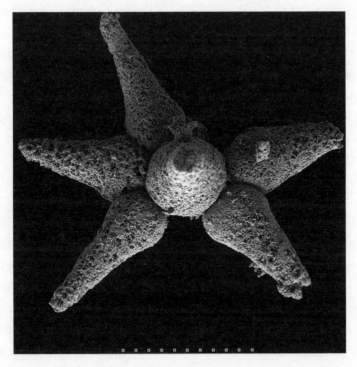

Figure 8. A small shelly fossil: a single chancellorid sclerite, lower Cambrian, Comley, Shropshire. Total length of scale bar is 0.5 mm.

Figure 9. (A) The chancellorid-like animal *Nidelric pugio*, lower Cambrian, Chengjiang Biota, Yunnan Province, China. Horizontal field of view for main image is 6.6 cm. (B) The spines of *Nidelric pugio* embedded in the body. Horizontal field of view is 1 cm.

Or rather it is described as 'chancelloriid-like', because the relations of these animals even to each other are still enigmatic.[16] It has small blade-like sclerites covering its body, inserting into the external 'leathery' skin, and seemingly designed to prevent predators from attacking its softer tissues. The animal appears to have been a hollow, balloon-shaped bag about 10 centimetres long, perhaps with an opening at the 'top'. It did not have bilateral symmetry like a worm or mollusc—a single plane through the animal about which either side is a mirror reflection of the other—but instead appears to have been radial, like a jellyfish or coral. This is where the chancelloriid plot thickens, because the sclerites suggest *Nidelric* is related to animals with bilateral symmetry. Hence, the simpler radial morphology in this case might have evolved from a more complex ancestor. The Cambrian thus emerges as a time of major evolutionary experiments in basic body plans, with the skeletons falling into place accordingly.

Seemingly seafloor dwellers, chancelloriids may have evolved for filter-feeding organic particles directly from the sea, adapting the strategy that is used by sponges. They had no need for locomotion, being protected by their spiny sclerites, and no need for organs either. These animals are sometimes referred to as 'cactus-like' in shape, and they have been variously thought of as relatives of sponges, or as having affinities with a range of other enigmatic Cambrian marine animals. The even stranger *Wiwaxia* is one of these possible relatives, and it developed its very own, and very distinctive, type of skeleton.

Wiwaxia's exoskeleton is also made up of many sclerites, though these appear not to have been biomineralized, instead being made of a tough organic material. As a result, when found disarticulated, one can class them as 'small carbonaceous fossils', just to add yet one more 'catch-all' taxonomic category.[17] *Wiwaxia* occurs in Cambrian rocks from China to British Columbia, its outlandish appearance resembling a punk-style hair-cut from the late 1970s. *Wiwaxia*'s sclerites come in different overlapping forms, which together covered the whole upper surface of the animal. Most of these were oval, though those that were in contact with the seabed were banana-shaped sclerites, while atop the animal were two rows of very long spines. Unlike the sedentary chancelloriids, *Wiwaxia* was mobile, and may have literally scraped a living by feeding on organic material at the seabed. It was obviously a successful life strategy, and other animals with exoskeletons—brachiopods—took advantage by hitching a ride on its spines as seemingly harmless passengers. What was *Wiwaxia?* Some scientists considered it as a kind of scale-covered worm, whilst others suggested that its 'rasping' mouth is reminiscent of that of a mollusc. Most people now seem to think that, amid the taxonomic fluidity of those times, it was at least halfway towards being a mollusc.

There is *Microdictyon*, too. Once known just from disarticulated oval plates with net-like surface patterns, these sclerites have now been found articulated, in the form of the exoskeleton of an early Cambrian lobopod, a kind of 'caterpillar-shaped' animal related to living velvet worms. They have revealed a further innovation too: fossilized plates of *Microdictyon*, stuck together, show the animal in the act of shedding its skin, with the new skeleton forming beneath the older one being moulted.

SSFs therefore record a multitude of different animals evolving in the early Cambrian seas and give us a sense of the beginnings of a rapidly evolving skeleton-based biosphere from some 530 million years ago. Then, about 520 million years ago a new group of animals appear in rocks worldwide. These had skeletons that covered the entire upper body with some *serious* biomineralized armour-plating, so clever in its construction that, in one form or another, it continues to the present day. The arthropods were on the rise.

Rise of the Planet of the Arthropods

Arthropods are not the largest of organisms, but they more than compensate by the diversity of body forms they have produced, their sheer numerical abundance, and their ability to live just about anywhere on the surface of the Earth. Arthropods were the first animals to colonize the air, and in this domain they can reach heights of over 5 kilometres. On the land they survive from hot springs to arid deserts. And in the seas they are found from the shoreline to the abyssal depths. Their diversity is astonishing. The biologist J.B.S. Haldane used to say that 'God has an inordinate fondness for beetles'—with 450 000 species so far recognized. That's impressive, but there are about twice as many insect species in total, when all are counted together. Then add crustaceans (such as crabs and lobsters), chelicerates (scorpions, spiders)—there are at least 45 700 species of spider—myriapods (centipedes and millipedes), and the numbers soon add up most impressively. By comparison, there are just 92 species of cetaceans (dolphins, porpoises, and whales), and 446 primate species (of which, the 7.3 billion humans account for well over 99.5% of the numerical abundance).

The arthropod abundance can be traced back almost to the beginnings of the complex marine ecosystems that have characterized Earth's oceans for at least 520 million years, and arthropod skeletons have been pivotal to building that complexity. The change is recorded in the fossil record by the sudden appearance of trilobites in early Cambrian strata from Greenland to Australia. This suddenness—which is likely a measure of the imperfection of the fossil record—is almost certainly due to their tough, biomineralized exoskeletons.

Arthropods, like other triploblast animals with bilateral symmetry in which the first opening formed in the embryo is the mouth (these animals are termed protostomes), make their exoskeleton from the outermost of the three layers of tissue that characterize their bodies, the ectoderm. In this, they differ from the deuterostomes (those animals whose first opening in the embryo is the anus), which make their skeletons from the middle layer of tissue, called the mesoderm. The latter animals include vertebrates like humans, but also, perhaps surprisingly, sea urchins (although the sea urchin skeleton may look external it is actually an endoskeleton, formed from the mesodermal tissues). In arthropods, the exoskeleton forms as a tough layer of chitin produced by a layer of epithelial cells within the ectododerm. Chitin is an organic polymer based on a glucose-like molecule, though laced with nitrogen (an N-acetylglucosamine, to be precise), the chemical bonding producing tiny crystalline fibrils. Chitin turns up widely across the kingdom of life, among both vertebrate and invertebrate animals, and within fungi too. It is pliable enough to allow the arthropod body to flex, but where more rigidity and protection is needed, for example in the fang-like mouthparts of spiders, or on the dorsal armour-plating of trilobites, the chitin is impregnated with a tough protein layer called sclerotin, or it is biomineralized with calcite.

The exoskeleton of arthropods is sufficiently flexible to allow their jointed limbs—the literal meaning of the word arthropod—to be very adaptable. The exoskeleton between these joints supports the soft tissues, and gives protection, whilst the joints provide a high degree of flexibility. It is an anatomical system that has proved remarkably successful. A visiting alien species might well christen the Earth as 'the planet of the arthropods'. Indeed, given the success of this body plan on planet Earth, that visiting alien might well look like an arthropod.

The geological appearance of trilobites in the fossil record is sudden, but there are hints of some kind of precursor. Trilobites are known to have made trace fossils called *Rusophycus*, which are oval depressions where they rested on the sea floor. *Rusophycus*-like impressions have been found in Cambrian strata that are a few million years older than the earliest known rocks that contain trilobite skeletons. This early *Rusophycus* may represent traces made by earlier arthropods related to the trilobites, albeit ones possessing an unmineralized exoskeleton that did not fossilize. The trilobites subsequently emerged in large numbers to become the first globally distributed large, mobile animals with a hard exoskeleton. Their abundance has been magnified because many of the fossils found are moulted and cast-off skeleton parts. Even allowing for this, once trilobites appeared, they quickly went on to form an important part of Cambrian ecosystems.

Their body armour gives us a first inkling of the remarkable tenacity of that skeleton. Trilobite armour was designed to withstand attack from predators, but is also a marvel of engineering that allowed complex articulated movement, such that some trilobites could roll up into a ball—presumably to avoid predators or other environmental stresses— whilst others would swim high into the plankton. Trilobed, with a large shield that covered the head, a body made up of a series of articulating plates, and a similarly modelled tail that often terminated in a spine, trilobites were built for survival. Beneath the skeleton, and sometimes preserved, was a segmented body with most segments bearing a pair of two-branched appendages—one branch for crawling, and the other housing gills for respiration. With excellent vision, and sensory antennae at the front, trilobites could see and feel their way around the ocean world of the Palaeozoic, and they are one of the defining fossils of that geological era.

Of all the components of the trilobite exoskeleton, perhaps it is the eyes that are the most magnificent invention of their evolution. In trilobites, and many other arthropods, these are compound, each being made up of multiple lenses, sometimes numbering in the thousands (Figure 10). In trilobites the lenses were mineralized with calcite, and in some beautiful examples of preservation, where each lens was made of a single crystal of calcite, it is still possible to shine light through those ancient eyes.

Such eyes served the Early Ordovician trilobite *Carolinites genacinaca* particularly well. It was a swimming trilobite that appears to have lived in the upper reaches of the water column, within the zone penetrated by light. It had a remarkably wide geographical range, occurring across the whole of the ancient tropics, and being found from North America to Australia. Its compound eyes were gigantic, giving it all-round vision to front and rear, and from above and below. Indeed, many reconstructions of *Carolinites* show it swimming on its back, using paddle-like limbs to propel it forward. Alas, the limbs of *Carolinites* itself are conjecture, having not left any record (or, at least, none has yet been discovered).

Trilobite eyes are wonderful, but some features of the trilobite skeleton are just deeply puzzling. For instance, many trilobites had different kinds

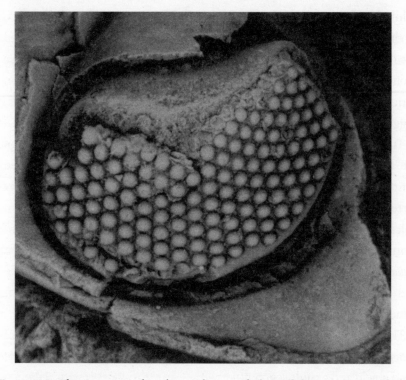

Figure 10. The compound arthropod eye of the trilobite *Ananaspis* from the Silurian Annascaul Formation, Dingle, Ireland. Horizontal field of view is 0.84 cm.

of thorn- or spine-like projections. Some of these spines simply projected sharply upwards from the exoskeleton, and these may be sensibly regarded as defensive, rather like (though not as numerous as) a hedgehog's spines. Others, like the spines that commonly extended from the head backwards along the side of the body, may also have been protective, by sticking up like lances when the trilobite enrolled, or may have helped the trilobite when in motion to achieve lift-off from the sea floor. Some trilobites had impressive single, or paired, or trident-like spines pointing forwards from the head, like the trilobite equivalent of a unicorn or a stag. These seem less obviously defensive or—as they were immobile—offensive. That doyen of contemporary trilobite study, Richard Fortey, has suggested,[18] they may have been a way of impressing trilobites of the opposite sex, in the manner of the tail of a male peacock, or the horns of a staghorn beetle. Another distinctive enigma is a broad flat fringe, covered in rows of pits, around the headshield of a distinctive group of trilobites, the trinucleids, which were common in the Ordovician Period. As these trilobites were eyeless, and presumably blind, it has been suggested that they might have served as some kind of sensory apparatus.[19]

The roughly 20 000 described trilobite species may be just a fraction of the arthropod story, because trilobites are serendipitously preserved simply because they have hard, mineralized skeletons. To view a fuller picture of the amazing burst of arthropod diversity in the early Palaeozoic, one needs to find rocks that have preserved not just the hard biomineralized fossils, but the softer anatomical parts too. Such rocks preserve look-alike relatives of the trilobites, and a host of other amazing and bizarre arthropods, which have exoskeletons, to be sure, but unmineralized ones that, in normal circumstances, did not fossilize.

Those look-alike relatives of the trilobites have a striking array of morphologies, though all built around the general plan of a semielliptical head shield and a multisegmented body. There is the marvellously named *Cindarella*, an early Cambrian animal that did not wear glass slippers, but possessed antennae emerging from the front of its head shield and a pair of stalked eyes emerging from below. Each of its eyes, which were compound like those of trilobites, bees, and spiders, had more than 2000 lenses, a marvel of microengineering that meant that *Cindarella* likely had little trouble detecting food, or potential predators, in the early Cambrian seas. And there is the 'see-through' trilobite-like *Naraoia*: through its exoskeleton its soft anatomy, including its peculiar 'tree-shaped' digestive tract, can be seen. *Cindarella* and *Naraoia* are known only in deposits that have exceptional fossil preservation of soft body tissues, like the Chengjiang Biota of south China.

Apart from trilobite-like animals, the early Palaeozoic seas were populated with many different arthropods. One such group is the stuff of real monsters—the anomalocaridids. These were distributed as far afield as what are now South China, Greenland, and western Canada, and were likely the apex predators of their day. Some reached over 1 metre in length. They are characterized by their large grasping frontal appendages that were greater in total length than many trilobites, and that had vicious spines for impaling prey. Their head flanked by pairs of stalked compound eyes, anomalocaridids must have been formidable predators, and the group probably survived well into the Devonian Period, more than 100 million years after its first appearance in the fossil record of the Cambrian.

In terms of numbers, arthropods dominated the earliest complex ecosystems of the Cambrian, just as they do in many marine ecosystems today. Of the 250 species described from the exceptionally preserved fossil biota of the early Cambrian Chengjiang deposits, one-third of the species are arthropods,[20] whilst numerically the most abundant animals within the whole assemblage were small arthropods called bradoriids, in which the body was almost wholly enclosed in a two-valved carapace. Evidence from the soft anatomy of these early arthropods, including the contents of their preserved stomachs, shows that these animals carved a living in many different marine settings as scavengers, as sweep feeders (literally sweeping up material to eat), as predators, and possibly even as filter-feeders.

Animals with internal skeletons were also present in these early complex Cambrian ecosystems, but as a very small component of the biota. These animals would spend the next 500 million years trying to catch up: they have had their success stories, as we shall see, but they still have not surpassed the arthropods in terms of their overall biodiversity.

The Enduring Ostracods

If a multitude of large and impressive arthropods filled the Cambrian seas, one group of very tiny and seemingly almost featureless arthropods was to outlast them all.

At the Aitsu Marine Station, Amakusa in Kysuhu, Southwest Japan, a young scientist called Gengo Tanaka collects marine ostracods by moonlight. He works late into the night, stooping at the edge of the quay, where he casts a small glass bottle on a piece of string into the sea. It has a pierced metal top, and inside it there is a small piece of pork sausage. He waits patiently and silently on the quay, then slowly lifts the bottle out of the water, carrying it to his laboratory nearby. There, in his lab, is a large glass bowl of seawater into which he has inserted more tasty morsels of sausage. He gently tips the contents of the glass bottle into the bowl, turns off the lights in the lab, and waits, to the sound of the whirring machines that keep the marine animals in their seawater tanks supplied with oxygen. In the dark, one can imagine it as Frankenstein's laboratory. As the machines whir, there is a sudden spark of light, and then another: inside the bowl, tiny, millimetre-long ostracods are feeding on sausage, and signalling to each other—not by electricity, but by light, bioluminescing as their ancestors may have done for tens or even hundreds of millions of years before them. This sausage-loving ostracod is called *Vargula hilgendorfii*.

Gengo has been carefully studying these particular ostracods for more than a decade now, practising with different kinds of food to see which ones they will prefer. Though tiny, the ostracods have a surprisingly long evolutionary heritage. In the mountains that form the deep core of the Japanese island of Kyushu there are very ancient rocks that date back more than 400 million years. There too, carefully uncovered by Gengo, are the petrified remains of ancient marine ostracods, distant relatives of those on the harbour at Aitsu Marine Station. By studying the living forms, Gengo hopes to find clues to how their distant ancestors lived, and how they have managed to survive across so much of Earth time.

Ostracods are aquatic arthropods, close relatives in the animal tree of life to crustaceans like crabs and lobsters. Typically never much more than a few millimetres long, they occasionally grow to gigantic proportions—for an ostracod, that is: the living *Gigantocypris* can be more than 1 centimetre long. Like other arthropods, ostracods moult several times before reaching maturity, each time preserving the intricate anatomy of their soft parts within their exoskeletons. These 'soft parts'—which in reality all have their own minuscule, delicate, unmineralized cuticle—include seven pairs of legs: four on the head, including those converted into a mandible for feeding; and three on their body for crawling and walking.

The very earliest ostracods evolved half a billion years ago in the sea. Over that immense passage of time they have learned to swim amongst the plankton, cruise the ocean abyssal plains, and to survive in hot springs, and can even drag a tiny ball of water with them to breathe amongst damp leaf litter. Unlike their more conspicuous arthropod cousins, the trilobites, they have also survived all five mass extinction events of the past 500 million years. How have they achieved this invincibility?

A tough but surprisingly simple outer armour is the key to their success. Instead of a complex, jointed mineralized exoskeleton, as the trilobites had, ostracods have tiny, paired shields that completely cover all of the complex unmineralized skeletal parts. It is a little like living in a reinforced and peripatetic suitcase (and for readers of Terry Pratchett, the inimitable Luggage might come to mind here). In some ways, these arthropods have adopted a general form of skeleton that the molluscs were to develop widely, only in diminutive and highly mobile form. It has proved to be a remarkably adaptable arrangement, with at least 65 000 different shapes and forms recorded from the fossil record. The extinct bradoriids evolved a similar-looking carapace, though the structure of their inner parts show that they are not closely related, and evolved this form independently. Bradoriids were successful in the Cambrian, but did not have anything like the geological longevity of the ostracods.

The exoskeleton of ostracods is mineralized with a hard and resistant mineral, calcite, like that of trilobites. The two 'shells' are hinged at the top, and can articulate: opening and shutting according to whether the animal is active or resting, and controlled by means of muscles. Ostracods can clasp these 'shells' tightly shut, allowing them to control the properties of sea- and freshwater next to their tissues, to withstand predators, or even to survive a (desiccating) long-distance journey out of water, carried, say, by birds today—and perhaps by many other animals in the past. Tiny as they are, ostracods are supreme survivors. In the biological catastrophe of the Permian–Triassic boundary, 252 million years ago, some 95% of marine species were exterminated—including the 270-million-year old lineage of the trilobites. It is the ostracods, though, that feature amongst the first animals found in the rock strata above the mass extinction event. If one was to devise an alien animal to survive under harsh conditions on some distant and catastrophe-prone planet, the ostracod model might be near the top of the list.

The Exoskeleton Invasion of Land

Many arthropods that now colonize the land had their origins in the sea. Some of those groups, like the ostracods, remain abundant in both settings, whilst others have left the sea forever.

A simple terrestrial biosphere, dominated by microbial life, may have evolved deep in the Precambrian over 2 billion years ago. But the development of a complex terrestrial biosphere of interacting plants, fungi, microbes, and animals is viewed as one of the fundamental transitions in the history of the biosphere. This began to form about 460 million years ago, when mosses gained a bridgehead on land. The first animals seem to have followed much later—though this is still debated—about 425 million years ago. Arthropods were in the vanguard of this animal invasion, and perhaps 50 million years ahead of the first endoskeleton invasions of land. Once again, as with the Cambrian biota, animals with exoskeletons staked their claims well ahead of those with endoskeletons. Why then were arthropods able to make this transition so early?

The earliest animals to find their way onto land seem to be more ancient relatives of that Carboniferous giant, *Arthropleura*. *Pneumodesmus newmani* is a name give-away. The 'newmani' part is simple. It is named after Mike Newman, the amateur palaeontologist who first discovered it in the 423 million-year-old rocks of Cowie Harbour, near Stonehaven in eastern Scotland. In fact, the rocks at Cowie Harbour contain the fossils of three different early millipede species. The ancestors of these animals, though, have their origins in the seas, so the discovery in itself was not unequivocal. It was a particular feature of *Pneumodesmus* that clinched its 'land-lubber' origin. The 'Pneumo' part of the name means breathing, for the exoskeleton of this animal contains the telltale spiracles that fed its respiratory air-breathing system.[21] Only a small part of *Pneumodesmus* is preserved, but it is enough.

As well as the many invertebrates with exoskeletons living in the Silurian seas at the same time that *Pneumodesmus* was colonizing the land, there were also many marine animals with endoskeletons. Vertebrates were already present amongst the Cambrian biota, and fish had been evolving for several tens of millions of years by the Silurian. Why was it that they did not immediately follow the arthropods on to land?

Arthropods had some definite advantages here. Firstly, they were small, and the first land plants were also small. Forests did not really begin to develop until the Late Devonian, some 380 million years ago. Amongst the low foliage of Silurian and early Devonian vegetation, small millipedes and the ancestors of insects and spiders could easily lurk. Secondly, arthropod exoskeletons resist desiccation. Arthropods did need to keep near to water to keep their 'skins' damp, as these were buried under a protective envelope of exoskeleton cuticle. And thirdly, perhaps arthropods could rapidly adapt to eating vegetation, whereas Silurian animals with endoskeletons tended to be large and amongst the top predators, consuming the bodies of other animals: becoming vegetarian was not an easy option for carnivores like the metre-long Silurian fish *Megamastax amblyodus*.[22]

One might think, then, that the first terrestrial arthropods that colonized the land had it easy, entering a new landscape of tiny, low-lying moss forests lacking large vertebrate predators. But remember those nearly 46 000 modern spider species: arachnids had an ancient ancestry too, and were in the vanguard of colonizing the land. As they quickly learnt to hunt down other arthropods, an arms race commenced that continues today.

Some of these earliest land-living arachnids are found in a famous rock layer about 420 million years old by the small market town of Ludlow in the Welsh Borderland. This is the Ludlow Bone Bed. It is usually only a few centimetres thick, and it looks like a layer of gingerbread mixed with crushed beetles, sandwiched in between some perfectly ordinary sandstones. The gingerbread part is made of what look like unusually large sand grains that on closer examination turn out to be the short, thick scales of primitive fish, while the 'crushed beetles' are mostly crustacean fragments. A geological accident, it represents winnowed skeletal remains, washed together by waves and currents along an ancient shoreline. Swept in from the nearby land, exceedingly rarely, are the remains of other skeletons—this time of some of the first terrestrial predators.

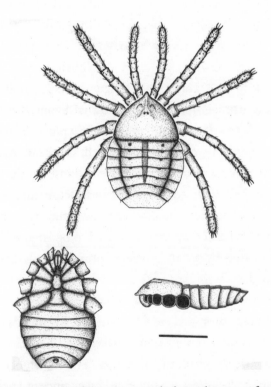

Figure 11. Reconstruction of the trigonotarbid *Eotarbus jerami* from the Silurian Ludlow Bone Bed Member, Downton Castle Sandstone Formation, Ludford Corner, Ludlow, Shropshire. Scale bar is 0.5 mm.

They are referred to a group called the trigonotarbids.[23] Tiny, usually a few millimetres in length, but with predatory knife-like fangs, the trigonotarbids are an extinct group of arachnids that were once widespread across the Earth: mind, even the fearsome-sounding *Gigantocharinus szatmaryi* was only 7 millimetres long! Trigonotarbids look superficially like spiders, but they are easily recognized as their abdomen is divided into a handful of 'tergites', a little like the subdivisions of a trilobite's thorax (Figure 11). In the Devonian rocks of New York State, trigonotarbids are found together with so-called 'rejectamenta', bits of other arthropods that they were eating, the indigestible bits being 'sicked up'. If these early terrestrial landscapes lacked vertebrates, the arthropods that occupied them already lived in a world of tooth and claw.

The Seashells

She sells seashells by the seashore. That tongue-twister was composed, it is said, with Mary Anning of Lyme Regis in mind, one of palaeontology's more remarkable characters and—in this case it is entirely appropriate to use the term—heroines. She did indeed sell, by the seashore, all kinds of fossil skeletons and shells, to keep herself and her family alive. But one would guess that precious little of her meagre profits came from selling simple 'sea shells' of the kind one associates with that label—the cockles and whelks and scallops that, in yet another traditional phrase, could be purchased by all and sundry, 'alive, alive-o', and not just as long-dead fossils.

For these particular skeleton designs were and are so widespread, so commonplace, that they occupy a kind of ever-present backcloth to all the more complex and exotic skeleton designs that vie for our attention and, in the case of Mary Anning's trade, for our money. Let us consider the two-valved shell, closing at a hinge, and containing and protecting the animal within. This has been an immensely successful lifestyle for the arthropods in the shape of the ostracods, although their minute size

44

means that this is a success that is invisible to most humans. At a larger scale—where one does not need a microscope or hand lens to observe the phenomenon—this is a design that was dominated by the brachiopods in the Palaeozoic Era. In the succeeding Mesozoic Era these animals vied for top spot with the up-and-coming molluscs, and then, in the modern times of the Cenozoic Era, the molluscs definitively took over. The brachiopods are still with us, but these days you generally have to look hard to find them.

The brachiopods used to be called 'lamp shells' because of the fancied resemblance of some of them to a kind of oil-lamp, but because the resemblance *was* fanciful, and because oil-lamps have long dropped out of fashion, the nickname is fading fast. Biologically, molecular studies show that in taxonomic affinity they are far from the arthropods, moderately far from the worms, and a little closer to the molluscs. They are mostly filter-feeders, the filtering structures themselves sometimes being supported on elaborate wing-like or spiral projections of calcium carbonate, all attached to and contained within the paired shells at the top and bottom of the animal, the shell at the bottom usually being the larger one that rested on the sea floor. Brachiopods are often attached to the sea floor or to some other solid object (that may be another brachiopod) by a fleshy stalk that comes out from near the hinge of the two valves.

Brachiopods can be found abundantly in Palaeozoic strata, to the extent that collectors often throw them away in disgust as they target more charismatic fossils such as the trilobites. This latter-day contempt is a tribute to just how successful they were, some of them carpeting parts of the Palaeozoic sea floor, pretty well to the exclusion of other animals.

The bivalved molluscs, which took a back seat in the early Palaeozoic but later boomed, can look superficially like a brachiopod. Indeed, one of the classic tests for undergraduate palaeontology students is to tell the two types of fossil apart, and the classic means of doing this is to use symmetry. In the brachiopods, the shells are dissimilar in size but each valve is typically symmetrical from side to side. The bivalves, by comparison, typically have two valves that are generally much the same in shape, but each valve is normally asymmetrical. It's a pretty good rule of thumb, though with occasional striking exceptions. One classic and abundant bivalve is the *Gryphaea*, or the 'devil's toenail', of Jurassic times, in which one valve is highly convex and enormously thickened and the other is thin and flat, like a lid. This mollusc was mimicking a brachiopod—and clearly doing so most successfully. This symmetry contrast is (usually) a good identification guide, but the real difference is that the brachiopod shells are at the top and bottom of the animal, while for the bivalves they are on the left and right side, the connection at the hinge line typically being strengthened by conspicuous interlocking tooth-and-socket structures.

Bivalved molluscs can come in a wide variety of shapes and sizes, and while many are filter-feeders, living on or near the sea floor, others have become deposit feeders, living buried in, and eating their way through, sediment below the sea floor. Yet others live buried within the sediment, but use siphons—fleshy tubes—that extend to the surface to bring to them water for filter-feeding, and, once filtered, to expel it. They can grow to impressive size. The largest living form, the giant clam of the Indian and South Pacific oceans, *Tridacna gigas*, can in good times—it is now vulnerable to extinction through over-harvesting by humans—be more than a metre across and weigh more than a couple of hundred kilograms, most of this being shell weight, and live for over a century.

How can it amass so much shell material? It helps to live in warm waters, from which calcium carbonate is easier to extract from solution. But *Tridacna gigas* has also borrowed a trick from the corals in having symbiotic single-celled dinoflagellate protists living within its tissues, and these, while obtaining shelter and nutrients from the giant clam, in turn provide most of the clam's nourishment through photosynthesis. With such super-charging, the shell grows enormous. As in many bivalves and brachiopods, the edges of the valves where they come together are thrown into spectacular zigzag-like folds, this being a good way to increase the length of the join along which water can come in and out, while at the same time helping to keep the shell almost closed when necessary, to exclude harmful objects or organisms. Among these harmful objects might, occasionally, be a diver's foot, although the sinister reputation of the giant clam for snapping shut around a diver's leg is a myth—the closure mechanism is too slow for it to be a significant peril.

The material of the shell itself, secreted by a tissue layer called the mantle, is a wonder, and in particular nacre, which many molluscs produce. This is the stuff of mother-of-pearl, which gives the insides of shells such as the abalone their beautiful iridescent sheen, and of pearls, too. Nacre scatters light so distinctively because it is made up of stacked tablet-shaped crystals of the aragonite form of calcium carbonate that have a thickness of about 500 millionths of a millimetre, which is similar to the wavelength of visible light. These tablets are embedded in an organic matrix in a 'bricks-and-mortar' pattern that is not only beautiful, but it is also tough, resistant to being pulled apart, compressed, or bent. Nacre forms only a part of the shell: outside of it there is typically a layer of prism-shaped calcium carbonate crystals, and outside of that there is a thin organic layer, the periostracum, that is a protective barrier between the shell and the water outside. The entire precision assembly of finely engineered microcrystals is somehow orchestrated by cells of the mantle and periostracum. Not all of this process is understood, and biomechanical engineers would love to know quite how this works, as they strive to analyse and replicate a process that the humble mollusc can do without thinking.

Pearls, with their almost supernatural allure to humans, are simply pieces of grit in this marvellous machine, which are coated by the animal in protective nacre. In real life, molluscs are good at excluding grit from their interiors—hence the rarity of pearls and their fabulous value in times past. These days, the long-suffering molluscs are farmed and implanted with mother-of-pearl seeds in the industrial-scale production of cultured pearls, and these are now a commonplace.

For a mollusc, two valves are clearly very good, but arguably one valve is even better—at least if one uses species diversity as a guide. The gastropods, or snails, which typically live in one shell in the form of a coiled tube, are also prodigious, with approaching 80 000 species having been recognized. They live in the oceans, along coastlines, in lakes and rivers and have conquered a realm that the bivalve molluscs have never entered, in being able to live on land, to the chagrin of many a gardener, where some—as those gardeners know all too well—have reverted to an almost shell-less form, the slug.

Gastropods can be harmless herbivores—outside of a well-tended garden, naturally—while some species are ferocious carnivores. For both lifestyles they make use of a dentition that in its way is as formidable as anything that a shark or a *Tyrannosaurus rex* ever possessed. This is the radula, a kind of mobile tongue covered in an array of tiny teeth that are used for rasping plant material or animal flesh. The teeth are not made of calcium carbonate, but of the complex organic polymer chitin, which may be tipped with crystals of iron oxide or silica for extra hardness. Carnivorous gastropods often target bivalves or other gastropods, and these victims can be recognized by the neat circular hole drilled into their shells by a combination of radula and acid secretions. The pretty-looking cone snails are predators of this kind, and they have modified their radula to be a poisoned harpoon, too, that injects venom into its prey before they feast on it.

Building a skeleton that can respond to such deadly attacks is not easy. One way is simply to thicken the shell, which is effective, but costly in terms of energy and construction materials. One gastropod, though, has developed a (so far) unique armour that literally has inbuilt iron cladding. This is *Crysomallon squamiferum*, a newly discovered species that lives in the deep sea, near to hydrothermal vents, where hot, volcanically derived water streams out of fractures in the sea floor. In this chemically charged environment, the *Crysomallon* gastropod builds a shell that includes an outer layer of crystals of greigite, an iron sulphide. The resulting armour provides an extra defensive layer both against a radula-bearing gastropod predator and from the crushing pressures generated by the claws of a local crab. This novel skeleton construction is effective enough to have made the researchers involved note that it may be of interest for human defence applications too.[24]

Such dramas can be not only seen, but sensed more directly if one possesses the perceptive fingertips of Geerat Vermeij, the California-based palaeontologist who has been completely blind since the age of three. Vermeij[25] has built a brilliant career out of analysing the nature of predation and predator avoidance in molluscs through the patterns revealed to him by touch. These patterns, of both successful and unsuccessful attacks, and shells both irreparably penetrated and part-penetrated and subsequently repaired, have given him a specific view of the history of life—especially where his tactile exploration of shells is extended back to the fossil forms of deep geological time, of the kind of cat-and-mouse game among shelled creatures that has been going on since *Cloudina* of late Precambrian times. There is, he says, an overall pattern through time of ever more resistant shells, and ever more aggressive weaponry, that has been a major driver of evolution. The arms race, to him, is real and can be sensed in more ways than just by sight. He has expressed this sobering viewpoint eloquently, including in some of the most beautifully written books about shells ever published.[26]

Another means of escape from the deadly radula is taken by some kinds of bivalve mollusc, notably the scallop. If an approaching predator is detected, they snap their valves shut, expelling water and propelling themselves a short distance through the water. It is a primitive form of jet propulsion. Other molluscs have taken this ability and developed it with a sophistication that is both breath-taking and beautiful. It was the cephalopod molluscs that evolved a fully functional jet engine.

Jet-Propelled Exoskeletons

It was not, therefore, the British pilot and engineer Sir Frank Whittle who was the first to design a jet. Nature made this propulsive leap forward in the Ordovician seas, well over 450 million years ago. There lived the ancestors of the modern *Nautilus*—not the submarine vessel steered through the oceans by Jules Verne's Captain Nemo, but a cephalopod, a relative of living squid and octopus. Of all living molluscs, like the squid and octopus, that use jet propulsion, *Nautilus* is the only one to retain its external shell, the others evolving them away long ago.

The external shell of *Nautilus* is made of the aragonite form of calcium carbonate. The shell is coiled, and internally it is subdivided into a series of chambers separated by internal walls. As the animal grows, it coils its shell, adding more chambers and dividing off preceding chambers. The latest chamber is called the living chamber, and it is within this chamber that the *Nautilus* does its jet-propelled magic. *Nautilus* is also a master of disguise, with the lower surface of its shell, when viewed below, being white to blend in to the sunlit waters above, while its upper-facing surface is striped, to blend in with the dark waters below. In this way the *Nautilus* avoids its predators.

In *Nautilus*, water is drawn into an area called the mantle cavity that lies below the main mass of its soft tissues: squid do the same, but they have a mantle cavity much larger than *Nautilus*, and can thus form a more powerful jet. In the mantle cavity the water is formed into a jet by powerful retractor and funnel muscles. The retractor muscles are attached to the shell, and they leave a muscle scar that is sometimes preserved in fossil material. The retractor and funnel muscles of *Nautilus* comprise less than 10% of the mass of the body. This ancient propulsive system is inefficient, though, compared to the 'fighter-jet engines' of modern squid, which are ten times more efficient in converting their jets of water to propulsive speed, and which can sometimes cause the animal to leap out of the water on flights of 50 metres to escape predators, as the only truly jet-propelled airborne animals apart from humans.

Nautilus is the only living mollusc with an external shell to use jet propulsion, but this form of propulsion was once widespread amongst its relatives, the ammonoids. From the Devonian Period onwards (and from about 410 million years ago), ammonoids evolved an extensive array of shapes and designs that survived the Late Devonian and end-Permian extinction events, but which succumbed to the end-Cretaceous extinction. Ammonoids range in size from a few centimetres to more than 2 metres in diameter. They evolved very rapidly, and as a result the fossils of their shells are used to recognize subdivisions of different ages in the rock record.

Ammonoids, and especially their Jurassic and Cretaceous representatives, the abundant and highly evolved ammonites, are spectacular and highly visible fossils. The specimens that are continually washed out of the crumbling shale cliffs of Whitby in Yorkshire were said, in local folklore, to be 'snakestones', the remains of a plague of snakes turned into stone by Hilda, the 7th-century abbess there. Hilda—as described by the Venerable Bede—was a woman of great energy, a fine administrator and teacher, but her gifts (for which she was later canonized) probably did not extend to the petrifaction of unwanted reptiles.

Much later, the 17th-century scientist Robert Hooke looked at ammonites more closely. Hooke has been described as 'England's Leonardo', a polymath who worked on the nature of gravity, on the physiology and morphology of animals and plants, the nature of air and of light, helped (with Sir Christopher Wren) rebuild London after its Great Fire, devised a better way to build watches—and found time to look at fossils too. He drew a variety of ammonites (beautifully—he was also a skilled draughtsman) and interpreted them as long-dead animals unlike any that live today, and hence extinct—yet similar to the living *Nautilus* (which he also drew for comparison)—that had been buried in stone and petrified by the action of mineralizing waters. It was an interpretation at odds with Biblical explanations then current, that these objects had been placed in the strata by the Creator, or were the remains of Noah's flood. Hooke's brilliant deductions were prescient, but his work as a whole was overshadowed by a quarrel later in life with the up-and-coming Isaac Newton—and Newton buried as much of Hooke's scientific legacy as he could. It was to be another century before savants such as the Comte de Buffon and later Baron Cuvier in France were to tread a similar path, to reach the same conclusions about these distinctive fossils.

Subsequently, as palaeontology developed as a science, ammonites became among the most avidly collected and studied fossils. There are some striking patterns on their elegant shells that have been much puzzled over. For instance, the pattern of the suture line, where the walls separating each successive chamber join with the main shell wall, shows a marked change through geological time. In early *Nautilus*-like forms, the wall is only gently curved, and the suture is simple. In later forms like the ammonites the wall develops complex fluted curvatures, and consequently the suture line shows a complex crenulated pattern. Why was this? It used to be generally thought that this was to help the whole ammonite resist the pressure of deep water, and so prevent implosion of the shell as the animal dived more deeply. The simple analogy used was comparing a simple sheet of paper with one that is folded several times: pushing at the edge of the former will easily bend the paper, while the folded paper will resist such lateral pressure. That seemed sensible and intuitive—but mathematical modelling suggested that the complex ammonite sutures were actually *less* good at resisting the pressure of deeper water than the simple sutures, but offered much better buoyancy control.[27] Simple intuition, therefore, is not always the best guide to interpreting the mechanical properties of complex skeletons.

The jet-propulsion systems of ammonites, squid, and other molluscs have been so successful over such a long period of Earth history that they have been recognized by the science of biomimetics, which is now literally following them. Submersible craft with a flexible body have been designed to copy their motion. Imagine the confusion, therefore, of a 14-metre-long colossal squid with its highly developed senses and brain, as a 200-metre-long squid-like submersible carrying humans passes by it silently in the water.

Skeleton of the Argonauts

Confusion of another sort was resolved by that most serendipitous of major 19th-century scientific figures, Jeanne Villepreux-Power. It concerned the nature of the argonauts. Not, in this case, the band of ancient Greek heroes, who accompanied Jason on his perilous voyage in the *Argo* to retrieve the Golden Fleece. Rather this is the little pelagic octopus *Argonauta* that possesses a delicate, rather beautiful coiled shell that looks quite similar to that of the *Nautilus* (and indeed the other name of the argonaut is the 'paper *Nautilus*').

But there are some very strange things about this molluscan argonaut. It can, for instance, leave this shell completely, unlike any other mollusc—and so sharp debates grew between savants as to whether it was its own shell, or one that it had appropriated, just as hermit crabs make use of discarded gastropod shells as mobile homes. And, between two of its tentacles, it has a sheet of soft tissue. Aristotle, at about 330 BC, described this tissue as a sail, by which the argonaut caught the wind to sail across the sea surface, rather like the ship of its mythical namesakes. This was a vision that persisted for more than two millennia, being reprised much later by Lord Byron and Alexander Pope in their poems, and by Jules Verne in his epic novel *20 000 Leagues under the Sea*.

Enter Jeanne Villepreux, the daughter of a mostly illiterate family in Juillac, in southern France. After a series of family difficulties—the death of a sister and mother, the arrival of a young stepmother—in 1812, at age 18, she made the 480-kilometre journey to Paris, on foot, with a flock of animals destined for the abattoir, and in the care of a cousin. The cousin attacked her partway through the journey, and she sought refuge in a gendarmerie, then a convent. Eventually she arrived in Paris alone, with no place to go to or prospect of work. As one of her biographers put it, chance can sometimes be merciful. A dressmaker took pity on her, and took her on as a seamstress. Unlike her siblings, she could at least read and write, and she learnt quickly, soon excelling at her new trade. A short few years later, when Princess Caroline of Sicily married the King's nephew, the Duc de Berry, it was the young Jeanne who designed the wedding dress. During the festivities, a young English businessman,

James Power, saw her, and fell in love. They married—and she started a new life as a rich émigré wife in Sicily.

Jeanne Villepreux-Power, as she now was, did not spend her life doing the social rounds, but instead became fascinated by the natural history of the island, and particularly the marine life. To get closer to the sea creatures, she needed some means of studying them at close quarters. To do this, she designed and had built a glass box filled with seawater—and thus became the inventor of the aquarium. She built three of these structures, two of which were placed within the sea, and one on land. Among the creatures she studied were the argonauts, having heard of the scientific dispute as to the nature of their shells. They were common in the seas around Sicily, and she thought herself well placed to get to the truth. Her confidence was well placed. She collected argonaut eggs, hatched them, and day by day, observed what happened. After a few days, tiny shells appeared—so clearly these were made by the organisms, and were not borrowed or stolen. And she saw that the delicate membranes that Aristotle took to be sails were the secretory organs for shell material, and could be used to repair as well as to build these delicate shells.

Unlike other women scientists of the day, her achievements were recognized by the almost exclusively male-dominated scientific establishment, and she became a member of not one but 16 scientific academies across Europe. Even the formidable Richard Owen, inventor of the Dinosauria and a man with notoriously sharp elbows when it came to jostling for academic position, heaped praise upon her.

The mobile shells of the argonauts are even more versatile than Aristotle had imagined. They are used as intermittent shelter, and for protection of the eggs. The animals, as they come up to the sea surface, also trap bubbles of air with their shells that they then use to regulate their buoyancy when diving deeper. Is this shell, therefore, a skeleton, a home, or a tool? Nature does not always abide by easy categories, and this problem sharpens as we consider the sophisticated constructions of the long-extinct graptolites.

The Building Trade

Fossils can be puzzling things, and their study has often been a succession of mysteries, and the resolution of these mysteries, discovery by discovery, each time brings the history of our planet a little better into focus. Some fossils have had a particularly picaresque journey through these interpretive shadowlands. Take the graptolites, for instance, given their name by no less than Linnaeus, back in the early 18th century. That father of biological taxonomy did give them their name, as *Graptolithus*, meaning 'writing on the rock' because of their hieroglyph-like appearance, but regarded them as 'false fossils', the result of some kind of inorganic process. That was a cautious and not unreasonable interpretation, given that some kinds of entirely non-biological crystal growths can take on linear or branching patterns.

By the 18th century, though, more specimens had been collected. Closer examination showed these strange objects to have an essentially tubular structure, as a series of tubes made of some kind of resistant organic substance, open at one end and joined to a main branch-like tube at the other (Figure 12). That they had some kind of biological origin seemed now clear—but what were they? Some mooted them as 'fucoids'—that is seaweeds—to put them into the plant kingdom. Others compared them to corals, or to their tiny relatives, the hydroids, which can be seen (with a good magnifying glass) in ponds and streams. Whatever they were, they came to be terribly useful. That Scottish schoolteacher who was to become an eminent professor, Charles Lapworth, found them in the rocks of the Southern Uplands of Scotland, and realized that the distinctive, evolving shapes, of what he saw as some kind of ancient plankton, could be used to characterize a succession of different time zones in Ordovician and Silurian strata. Mysterious or not, they were clearly helpful to geologists as a means of dating ancient rock strata.

In the 1930s the Polish palaeontologist Roman Kozłowski stumbled across some graptolite specimens, perfectly preserved—almost like biological specimens—within that pure silica rock, chert. He made a minute

Figure 12. *Spirograptus* graptolites preserved flattened in Silurian mudrock from the Czech Republic. Scale bar shows cm.

study of them, dissolving them out of the rock with dangerous, and highly toxic, hydrofluoric acid, and taking thin slices of them for study with a microscope, and then faithfully reconstructing their three-dimensional shape. He realized that the nearest comparison was not with plants, or hydroids, but with an obscure group of marine creatures called the pterobranch hemichordates. Pterobranchs are tiny, colonial tube-dwelling filter-feeders that live on the sea floor. At first glance their tubes do not look much like the precise and elaborate constructions of the planktonic graptolites, being much more untidily arranged. But their detailed structure, being made of a series of rings, was very similar.

It was a breakthrough, but one that very nearly never saw the light of day. As Kozłowski was preparing to publish the work, the Second World War began, and his ordered life was turned upside down. When war broke out, he hastily sent negatives of his photographs to Paris, and stored his manuscript in a university basement, only for the building to be destroyed as the Nazis advanced. Somehow, he later found the scattered pages in the rubble. A few years later, as the Warsaw uprising began, he hid the manuscript in some heating pipes. Again, the building, like much of the city, was destroyed, though yet again both he and his manuscript miraculously survived, for him to eventually publish his findings, with the help of the retrieved negatives, after the war.

A palaeontological problem solved?—to an extent, certainly. In the following decades, most palaeontologists accepted Kozłowski's interpretation of graptolites as pterobranch relatives. A standard reconstruction showed the complex tube systems enveloped in, and secreted by, a surrounding layer of soft tissue, in the way that internal skeletons such as our bones are secreted. It seemed the obvious way to form such elaborate structures.

There was a nagging problem. The tubes of modern pterobranchs are not like bones or shells, secreted by a tissue layer. They are *constructed* by the organism, which actively builds them, ring by ring, much like termites build their nest, or a spider constructs its web. Could the graptolites have done the same? This seemed doubtful, for comparing the simple, untidily arranged tubes of the pterobranchs and the precisely engineered constructions of the graptolites was like comparing a medieval cart with a racing car. Tiny pterobranch-like organisms, with no real brain and a simple nervous system, were not thought capable of actively—and cooperatively—building such elaborate skeletal architecture.

Then, in the late 1970s, the doyen of studies into graptolites (who lived a double life as one of the country's best known pike fishermen) Barrie Rickards of Cambridge University and his PhD student, Peter Crowther, were looking at finely preserved graptolites using what was then a revolutionary new machine—a scanning electron microscope, which can reveal microscopic surface details beautifully. They noticed that the outer layer of the tube was formed of a mesh of what looked like irregularly criss-crossing microscopic bandages, which covered the rings of the tubes just as a layer of plaster covers a brick wall. They realized that such a structure could not sensibly form within an enveloping layer of soft tissue, but must have been actively 'trowelled' on to the wall of the tube by the 'plasterer'—the organism that lived inside the tube. And, by extension, the 'plasterer' must have previously been the 'bricklayer' to construct the tube itself—just as modern pterobranchs do, but to a much more sophisticated blueprint.

It was another breakthrough, at first doubted, for some of the structures that must have been 'built' cooperatively in this way are astonishingly complex, equalling if not surpassing in some respects the feats of the termites. Now, though, this interpretation is widely accepted. But it is a breakthrough that further stretches and blurs the definition of what a skeleton is, as the graptolite tube is made of hard supportive material that is external to but intimately linked with the animal, just like a mollusc shell—so is it now an exoskeleton? For, equally, it seems to be clearly a built and designed structure like the termite nests, or like another example that is sometimes fossilized, the acorn-shaped nests built by solitary wasps in some volcanic soils, put together out of carefully selected grains of pumice. So maybe, viewed that way, the preserved graptolite is not a skeleton at all, but a home. In some ways, the graptolites might be considered as an ancient forerunner of the question one has in considering skeleton-like structures built and designed by humans. Since they were highly mobile, they might be akin to the automobiles—or rather the submarines—that we now build, as forms of mobile self-built exoskeleton. Or perhaps, simply, we should not always seek to impose too strict a classification of what is or what is not a skeleton, but just wonder at the capacities of the natural world to spring wonderful surprises.

The Slow Road to Success

One thing is very notable in the exceptionally preserved animals of the Cambrian—that animals with exoskeletons, especially arthropods, ruled the seas. Those seas were also populated with the chancelloriids, with their spiny external skins, the wiwaxiids with their punk-style scale patterns, and many other groups of Cambrian animals with exoskeletons that together include early versions of molluscs, lobopods, brachiopods, priapulid worms, and the graptolites.

But where are the Cambrian animals with skeletons on the inside? Among the animals of the Chengjiang biota was the tiny finned *Myllokunmingia*, a jawless fish. Much, much later, the descendants of these early vertebrates were to reach 30 metres in length, swimming from the warm tropics to the cold polar waters to feed on countless arthropod krill. But for now, *Myllokunmingia* hid away in the shadows, avoiding the lethally well-armed arthropods in the water column above. It was an early representative of one of Earth's slowest-burning success stories. But when success finally came, it was to take remarkable form.

3

A SHELL ON THE INSIDE

In the 1966 cult film *One Million Years BC*, dinosaurs do battle with Raquel Welch and fellow humans of the Stone Age. Posters advertising the film stated that 'This was the way it was', ignoring that dinosaurs had been extinct some 65 million years, with the first *Homo sapiens* only walking the planet about 300 000 years ago. No matter. As the celebrated animator Ray Harryhausen once noted, 'professors probably don't go to see these kinds of movies'. Harryhausen was the animator of dinosaurs and skeletons par excellence. His models of the seven fighting human skeletons attacking Jason and the Argonauts in the 1963 movie of that name became cinematic legend. In reality the skeletons were only 10 inches high, and one of them had been recycled from an earlier movie about Sinbad. But, in Harryhausen's hands, they came back to impressively sinister life.

The sight of walking skeletons would be unlikely to perturb any arthropods in the audience, though, given that these animals wear their mobile skeletons on the outside from the outset. What might puzzle arthropods more—if they were prone to speculate on such things—is a conundrum: why would any animals develop a skeleton on the inside— an endoskeleton—given that such an arrangement gives no obvious protection against attack? That very act of building an internal rather than external skeleton, from such a perspective, might seem to simply offer a readily obtainable meal to a passing predator. Indeed, in the earliest preserved animal-based ecosystems of 500 million years ago, animals with internal skeletons were very much in the minority—and they were small too, as if to hide from the much larger arthropod predators such as *Anomalocaris*. In a sense, the situation is not so different today. Whilst there are many more than a million documented exoskeleton-clad invertebrate species, with insects alone accounting for about 1 million of those, there are a little less than 70 000 vertebrate species with their internal skeletons, including all of the mammals, birds, amphibians, reptiles, and fish. To a second approximation, our planet remains the planet of the arthropods.[28]

The answer that one might give to a puzzled, and philosophical, arthropod[29] is that there are some real advantages to building an internal skeleton. Where food is plentiful it allows an animal to grow very large, unconstrained by the need to keep moulting a skeleton. And to grow big can be a means of predator avoidance. Elephants eat plants, and lions eat meat, but lions do not eat adult elephants because elephants are just too big and powerful to contend with. So size can provide an advantage. Having a skeleton on the inside also means that the outside of the animal can better sense the environment around it, while it does not preclude other forms of armament, as developed today in animals such as living hedgehogs and armadillos, and in the fossil record by armoured dinosaurs and, yet earlier, by armoured fish.

Nevertheless, if animals with internal skeletons were one day to develop to the scale of a *Brontosaurus*, a mammoth, and a whale, their earliest origins were much less impressive. The distant ancestors of these behemoths were a barely noticeable, and only recently discovered, part of the Cambrian marine ecosystem of half a billion years ago. What used to be widely regarded as the earliest fish, though, came from much younger rocks. They were amongst the most beautiful and bizarre fossils that palaeontologists have ever sought amongst the ancient rocks of the world.

The Armoured Fish

When, in 1844, that eminent and immensely productive scientist Louis Agassiz introduced his monograph on the fossil fishes of the Old Red

Sandstone, the distinctive succession that makes up most of the Devonian strata of Britain, he opened by paying elegant tribute to those who had helped him. He thanked the British Association for the Advancement of Science for inviting him to do this study, and for funding the travel for him. He thanked the British geological pioneers, too, for their perseverance and zeal in making discoveries in strata previously thought barren. The usual suspects were there among the eulogies—Adam Sedgwick and Roderick Murchison, those giants of Victorian geology who had carved enormous chunks of the time scale out of the older and more gnarled rocks of Britain. But there were special thanks, too, for a considerably less aristocratic personage—Hugh Miller, a Scottish stonemason turned accountant, a self-taught scientist, and popular science writer of the day. A few years earlier, Miller had written his own semi-poetic, semi-scientific account of the Old Red Sandstone, including images of the fantastical fish that he had excavated from those unforgiving strata (Figure 13).

What extraordinary fish they were! Agassiz himself waxed lyrical in his introductory pages, comparing them to the discoveries made, a quarter of a century earlier, of ichthyosaurs and plesiosaurs (not least by that other emphatically non-aristocratic and self-taught pioneer, Mary Anning of Lyme Regis, who Agassiz was to go on to honour by naming two new fossil fish species after). Clad in intricately patterned and lustrous armour, with heads shaped like shovels or rockets or rectangular boxes, some with bony wing-like extensions, these fossils were both charismatic and puzzling, Were they tortoises, or crustaceans, or even some kind of beetle? Agassiz affirmed them as the earliest fish then known, and described them with characteristic clarity and thoroughness.

These beautiful fossils, though, seem to break the vertebrate rules. Our bones are emphatically on the inside—and yet here were these fish with hard carapaces, seemingly playing the same game as the arthropods. Indeed, they have been generally termed ostracoderms (meaning 'shell skinned'). These vertebrate creatures could, fresh out of the sea,

Pterichthyodes milleri
Middle Devonian
Achanarras, Caithness, Scotland
16139

Figure 13. *Pterichthyodes milleri*, the bizarre armoured fossil fish of Devonian age from the Old Red Sandstone of Scotland that caught the imagination of both the 19th-century stonemason and self-taught geologist Hugh Miller and the famous Swiss scientist Louis Agassiz, who honoured Miller by naming this species after him. *Pterichthyodes milleri* grew to about 30 cm long.

have auditioned for Ray Harryhausen.[30] Examine that armour closely, and there is another shock. The armour plating, whether as thousands of button-like scales or sheets of armour, is made of the mineral apatite, chemically calcium phosphate, the same as our bones. But structurally this early fish armour is not like our bones, which are secreted by special bone cells. Here, the lustrous surface is formed of hard enamel, below which is dentine, with scattered pulp cavities. Structurally, these are *teeth*, surrounding the animal on the outside, and designed for protection and not for biting. Even Salvador Dali's imagination would not have thought up an arrangement like this. So—where did these strange fish come from, and how did they then lead to the more familiar vertebrates of our times, which typically carry their bones deep within their soft tissues?

The Old Red Sandstone is sometimes popularly called the Age of Fish. It is some 150 million years more recent than the Cambrian Period with its explosion of multicellular, and often skeletonized, life. The strata of the Old Sandstone are often quite as red as the name suggests, as a walk through the countryside of, say, Herefordshire or the Midland Valley of Scotland will show you: here, ploughing exposes the deep ruddy hues of the soils derived from the red rocks beneath. These rocks are red because they originated on the semi-arid landscape of the Old Red Sandstone continent, where the oxygen in the air simply rusted the sediments as they accumulated, mostly on river floodplains. The Devonian rivers were home to the array of bony fish that were among the first invaders of the land, as our planet developed a terrestrial biosphere to go with the marine biosphere it had harboured for more than 3 billion years: the fish that, petrified, were later to captivate Miller, Agassiz, and their successors.

Did vertebrates then *originate* on land, much later than all of the other main groups of animals? One could be forgiven for thinking so, amid this first cornucopia of primitive-looking fish that was hauled out of the rocks. However, the patient and persistent hammers of succeeding generations of geologists put that idea to rest. Fish did turn up in older rocks, with the Devonian bonanza turning out to have started within the preceding Silurian Period—and with fish being subsequently found in both marine and non-marine strata of the Devonian. Before that, there was not so much to be found, but—following a good deal more hammering—not quite nothing at all.

Late in the 19th century, fragments of a bony fish-like carapace were turned up in an older rock formation, the Harding Sandstone, in and around a quarry 'about one mile north-west of the state penitentiary' of Canyon City, in Colorado. These rock strata were yet older, and substantially so, having formed midway through the preceding Ordovician Period. The man behind the discovery was the indefatigable Charles

Doolittle Walcott. At the beginning it was just a routine enquiry—Walcott had heard of a new locality with early Palaeozoic fossils from one of the local Geological Survey collectors. There was nothing unusual here in the slightest. It was just a supposed local extension of the distribution of these rocks—the bread and butter of geological mapping, nothing more. He asked for a few more fossils to be collected and sent to him.

It was in routinely examining these that he made the discovery. Among the fossil shell and coral fragments, he discerned what he thought were fragments of fish armour. It must have been one of those bombshell moments, even for a man as used to major discovery as Walcott was. He immediately asked for more specimens to be sent to him, and then made the trek himself to Canyon City to see and to collect the material for himself. His published description in 1892 was meticulous, including a complete list of all of the other fossils in these strata—all marine, and including several brachiopod and trilobite species—and detailed, including microscopic descriptions of the fish remains themselves. He was clearly anticipating questions as to the age of the rocks (are they really mid-Ordovician?) and the nature of the armour fragments (are they really fish?). These questions duly, and promptly, emerged from his surprised colleagues, but Walcott, having done his groundwork thoroughly, was able to answer them, emphatically, in the affirmative—including noting the 'tooth-like' character of the fish skeleton. Waxing lyrical in turn, he noted that these fossils represented 'diminutive ancestors of the great fishes that at a later date swarmed in the Devonian sea and left their remains in the classic "Old Red Sandstone"'.

So, the fish record went back to Ordovician times. Occasionally, such ancient fragments were turned up elsewhere in rocks of this age. But these representatives of our distant ancestors were clearly rarities, a vanishingly small part of the early Palaeozoic marine fauna, just here and there inhabiting some sandy shoreline. Then, for a long while, the trail went cold. What might have been the *ancestors* of these ancestral forms?

The Earliest Vertebrates

There is another great jump of time in the vertebrate story that leapfrogs most of the Cambrian to land in its early part, amid one of the most celebrated of the fossil Lägerstatten, where organisms are petrified with extraordinary fidelity, allowing a glimpse of the true nature of ancient biological communities.

This is the 520-million-year-old Chengjiang Biota of southern China,[31] now legendary for the almost routine preservation of soft anatomy—eyes, skin, gut, muscles. It has yielded many spectacular fossils, of which one, *Myllokunmingia*, is something of a bit player, neither common nor outwardly spectacular. It is, indeed, the kind of fossil that one of the experts in early vertebrates, Philippe Janvier, calls a 'squashed slug'. Nevertheless, the features that remain suggest that it is the earliest known fish (Figure 14).

Myllokunmingia, when found, looked uncannily like the hypothetical 'ancestral vertebrate' pictured by palaeontologists as based on the modern lancelet, *Amphioxus*, a small marine animal that does not possess a vertebral column, but instead has its forerunner, a notochord, which is a cylindrical toughened rod running down its back. *Myllokunmingia* seemingly had a notochord, too, as well as dorsal and ventral fins and a tail—so it was almost certainly a chordate (a more general category than vertebrate) and perhaps it was a vertebrate too. It also had eyes—at least,

Figure 14. *Myllokunmingia fengjiaoa*, lower Cambrian, Chengjiang Biota, Yunnan Province, China. The head-end of the animal is to the right. Horizontal field of view is 4 cm.

if the current interpretation of two preserved circular structures on the head is correct—and gills, structures that would turn out to have profound importance for the subsequent evolution of the vertebrates, and blocks of muscle along its body, seen in the fossil specimens as zigzag patterns. These muscles must have connected to an internal skeleton of some sort—but, as this is not evident in the fossil, it was probably not hardened with mineral as our bones are, or as were present in the bony carapaces of the Devonian fish, but was more like the soft cartilaginous bones of a shark. All this suggests that *Myllokunmingia* was a swimmer—and perhaps was able to make powerful tail flips to avoid predation by some of the giant arthropods of the Cambrian seas. Whether it was a predator of smaller animals itself, or a filter-feeder, or mud-grubber, is still a mystery. Rare, small, obscure—*Myllokunmingia* was to set the pattern for vertebrates for the next 100 million years.

With the discovery of *Myllokunmingia*, it is now clear that the vertebrate lineage runs right back to the Cambrian explosion (other such 'squashed slugs' have been found in the Chengjiang and in other early Palaeozoic strata). However, the vertebrates mostly stayed either small and rare, or soft bodied (or more precisely 'soft skeletoned') until their prodigious radiation in late Silurian and Devonian times. While the external body-covering bone of the ostracoderms resembled that of our own teeth in microstructure, these organisms did not yet develop jaws and teeth as we understand them.

However, teeth of another kind were present in those seas. It is a pity, perhaps, that they were unconnected to any kind of skeleton. For this meant that they stayed one of the greatest puzzles—indeed, the greatest and most *useful* puzzles—in palaeontology for over a century.

The Strangest Teeth

Through most of the rest of the Cambrian and Ordovician periods, the seas teemed with trilobites and brachiopods, corals and nautiloids, their remains crowding in strata of that age. But in those strata, there are vanishingly few fossils that represent obvious, thoroughgoing fish. The vertebrates seemed to have kept the lowest of profiles among the exoskeleton-clad hordes—even when they tried on an exoskeleton themselves.

But—almost from the earliest days of palaeontology—small, terribly puzzling, fossils kept turning up in the rocks of Palaeozoic age. They were called conodonts. Sometimes they could be seen on a rock surface looked at through a binocular microscope. More commonly, when a limestone was dissolved in acid, dozens or even hundreds of these objects would appear at the bottom of the beaker, as the most beautiful and striking part of the insoluble residue. Some looked like complex hacksaw blades, just a millimetre or two long. Others were like tiny necklaces of elongated cones. Yet others were flat and knobbly, with a ridge running down their length. What were they? Some looked quite tooth like, while others looked anything but. They were made of calcium phosphate, the stuff of which our bones are made—but then, other groups of animals such as brachiopods use this material too. Also, teeth are usually attached to jaws and bodies. Whatever these objects were, in the rock they appeared jawless and bodiless.

Mystery or not, conodonts were most useful. They changed their shape over time, as the invisible (to palaeontologists) creatures that grew them evolved through much of the Palaeozoic Era. They are common, too, as the average lump of Palaeozoic strata, when dissolved in a beaker of acid, will typically yield some of these fossils. Hence, they are useful time markers for strata—provided, of course, that one has the expertise to recognize the differences between the many hundreds of species that have been identified.

So they carried on being widely used by pragmatic geologists, while those more inclined to ponder on their affinities pursued various hypotheses. Could they be worm teeth? The idea is not so crazy as it sounds, as some worms do indeed make arrays of 'teeth' called scolecodonts, which can also be found in Palaeozoic rocks. Scolecodonts are of similar size to conodonts, and some are of broadly similar shape, though they are made of a hard organic substance, not calcium phosphate. Or perhaps, others thought, they could be part of some arthropod. Or of some mollusc. Or of some plant. Or even of some creature not like anything living now; the palaeontologist Maurits Lindström in the 1970s hypothesized that the teeth-like elements surrounded the outside of some unknown animal for protection, as a kind of prehistoric and personalized barbed-wire fence.

Then came the 'eureka moment', the discovery in 1983 of the animal itself, residing in a museum tray in Edinburgh, together with swarms of fossil shrimps. And it was an animal too—and a vertebrate, though not of any kind now living. This first 'conodont animal' came from the 330-million-year-old rocks of the Carboniferous Granton Shrimp Bed, and it preserves large eyes, zigzag-shaped muscle blocks, and an eel-like body with caudal fins. At the front of the animal is an array of those long-mysterious conodonts assembled in the form they took in life, as arrays of some 15 'teeth' that were used for grasping and grinding food. It seems an almost phantasmogorically complex arrangement for catching and eating prey—but it clearly worked well for over 200 million years (Figures 15 and 16).

But, though conodonts may have evolved the first mouthparts that could process food in a vertebrate, they did not possess the first vertebrate teeth as we understand—and possess—them. A detailed analysis of the microstructure of conodonts conducted by Duncan Murdoch and Philip Donoghue, palaeontologists at Bristol University, showed conclusively that these 'teeth' had evolved independently of those in jawed fish, and thus also independently of those of amphibians, reptiles, birds, and humans, all of which have their ultimate origins in the bony external armour plating of the ancient fish that Hugh Miller, nearly two centuries ago now, hammered out of the Scottish rocky landscapes.

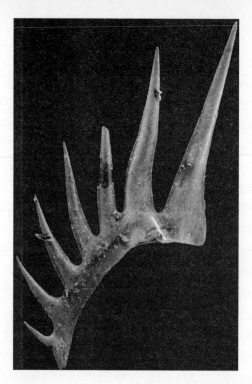

Figure 15. An 'S' element of the Carboniferous conodont *Idioprioniodus*. The element is about 1 mm long.

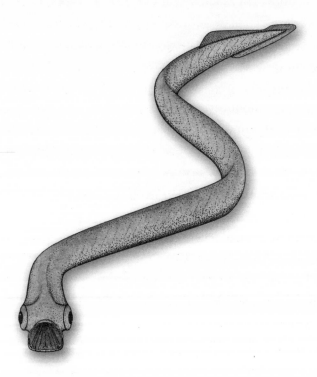

Figure 16. Anatomical reconstruction of a conodont animal, based primarily on specimens of *Clydagnathus windsorensis* from the Carboniferous, Granton, Edinburgh. The body was about 4 cm long.

Building Blocks of a Skeleton

In some of these earliest fish, a calcium phosphate skeleton formed—if slowly and haltingly, and at first as a kind of wrap-around tooth. What was special about this material?—and what led vertebrates to persist with it, instead of turning to that easy, readily available and tried and tested substance of choice, calcium carbonate, as so many other major animal groups did?

In those early days of skeletal experimentation, there were probably a few reasons for persisting with this material. Firstly, calcium phosphate, in the form of the mineral apatite, is significantly harder than the calcium carbonate minerals calcite and aragonite. On the ten-mineral Mohs' scale of hardness that all geologists learn (talc, the softest is 1, while the hardest, diamond is 10), calcite comes in at 3, while apatite is 5. Because the Mohs' scale is relative, apatite is in fact five times harder than calcite. That kind of hardness difference may have been useful to the clumsy, slow-moving early armoured fish, as they tried to evade becoming a meal for the mobile and aggressive arthropods of those days. Secondly, calcium phosphate is less soluble than calcium carbonate, which is handy under conditions where the ocean waters become more acidic and might threaten the integrity of a calcite skeleton (as such skeletons are being threatened today, with ocean acidification). Thirdly, phosphorus is a key element in the physiology of organisms: in animals it lies at the heart of the chemical energy release mechanisms that drive their activity. Possessing a storehouse of this material as part of the body's structure, to be called upon when needed, may have been the first function of this mineral substance in the animal.

In a vertebrate animal, one may speak of four main tissues that make up the skeleton. There is cartilage, a tough, elastic but non-mineralized tissue (and so one that, to the frustration of many palaeontologists, does not fossilize easily); a little of this material is all the earliest fish had as skeleton, mainly to support the gills and notochord (as have the hagfish and lampreys of today). There is dentine, where the calcium phosphate biocrystallizes on to a microscopic meshwork of collagen fibrils. There is enamel, a hard shiny substance with little or no organic framework. And there is bone, which is typically a complex 'living' skeletal material, which includes the bone cells that control mineralization, together with blood vessels and nerves; it is also a major biochemical factory in its own right, harbouring the internal bone marrow where blood cells are formed.

The dentine- and enamel-like early armour of ostracoderm fish formed, by bioprecipitation of hydroxyapatite, in the lower layers of the skin, and then emerged at the surface, much as teeth erupt through the skin today. Indeed our teeth are the remains of this external layer, now confined to the inside of the mouth. In the ostracoderms, this 'bony' armour typically solidly encased and protected the head region, behind which it was divided into thick scales, allowing protection to be combined with flexibility.

This early armour included layers that have been called 'acellular bone', but there was no true cellular bone, and no mineralized internal skeleton. These developments were, though, soon to emerge. In the fossil record they are linked with another skeletal innovation—the jaw.

Tooth and Jaw, but Not Yet Claw

Jaw jaw not war war, as Winston Churchill was wont to say. However, jaw and war—or at least aggression with lethal intent—all too often go together extremely well. The great white shark, *Carcharodon carcharias,* star of the 1970s monster movie *Jaws,* can grow to over 6 metres long. Its mouth can bite with a calculated force of 18 kilonewtons—perhaps a tenth of the force of the thrust of the engine of a jet fighter. Its ancient Miocene relative, *Carcharocles megalodon,* was a yet larger ocean predator, at 17 metres long with 12-centimetre teeth forming a fearsome array inside a pair of jaws that, fully agape, could accommodate—briefly, perhaps—a standing 6-foot human (Figure 17). (The experience may be safely recreated by visitors to Washington's Smithsonian Museum of Natural

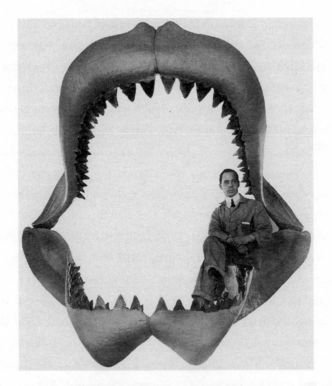

Figure 17. The jaws of megalodon.

History, who can have their photograph taken through the gape of a fine specimen of such jaws).

This is a scale of size and power that leaves the arthropods far behind. And jaws, of course, are not just used for biting prey. The baleen whales—and the 12-metre whale shark too—use them to filter plankton, in gigantic amounts, from the seawater. Elephants have massive grinding molars, to chew through their daily ration of more than a hundred kilograms of plants. Hummingbirds prolong their jaws, as beaks, to sip nectar in carefully controlled flight. And we humans now use jaws, of course, to talk of both war and peace. So where did jaws come from? The development of these skeletal structures was the key to much else that followed in the history of bony vertebrates, and so the 'origin-of-jaws' question has been one of the most closely studied in palaeontology.

The first clues to this were found amongst the plethora of Devonian fish that so caught the eye of Miller and Agassiz. The charismatic ostracoderms were jawless, like modern lampreys and hagfish. Ostracoderms are thought to have largely fed via a kind of suction exerted by their pharynx, pulling slowly into their mouth any prey that was too slow to get out of the way. This was linked to a major evolutionary innovation, in that ostracoderm gill slits began to be used only for breathing. In that model 'vertebrate ancestor', the lancelet *Amphioxus*, for instance, the gill slits are solely there for filter-feeding, with breathing carried out through the skin. This change was to have further consequences.

The menagerie of fish that were hauled out of the prolific Devonian strata were not all jawless. The armoured fish that most astounded Agassiz was one he called *Pterichthys*. Like a foot-long armoured box with a scaly tail, possessing two large bony 'arms', he called it the most bizarre thing in all of fishy creation, and said that he felt the same astonishment on seeing it that the great Baron Cuvier had expressed when examining a *Plesiosaurus* for the first time.[32] Agassiz thought that *Pterichthys* lacked jaws, like many of the other armoured fish he described. But however hard he peered at them, it was difficult to make out their critical anatomical details—these were squashed *armoured* slugs, a fragmented and compressed puzzle that was too difficult to be solved in those days.

More than half a century later, the Swedish palaeontologist Erik Stensiö painstakingly ground down fossil specimens of some related fossil fish, stopping each millimetre or so to fashion wax slices shaped from the fossil outlines seen on the ground-down end. The magnificent fossils were destroyed. But, slice by slice, their detailed internal structure was assembled as three-dimensional wax models. By this means, a new group of ancient fish emerged: these were the *placoderms*, with armoured carapaces outwardly similar to those of the ostracoderms—but with distinct, and well developed, jaws. Once recognized as such, their history was traced to well before the Devonian, with examples dating back to early in the Silurian. *Pterichthys* (now called *Pterichthyodes*) was just one example. A more impressive example of those days was *Dunkleosteus*— 2 metres long, streamlined, with a cruelly serrated jaw like something out of science fiction. It has been called the world's first vertebrate super-predator, able to take smaller predator fish as its own prey. How, though, does one make a jaw?

For the making of vertebrate jaws, it seems very likely that we have to go back to the gills, and in particular the gill arches—toughened structures that helped keep the gill slits open. In bony fish, the first of these arches seems to have evolved to become the jaw. This process is recapitulated in the development of the skeleton of a human foetus, where the first supporting arch of the pharynx, seen in the 4-week-old foetus, subsequently develops into the jaw—a strong clue to its evolutionary origins. It is from the same tissues connected with this first pharyngeal arch in the human foetus that the malleus and incus bones of the middle ear are also sourced, indicating a jaw-bone origin for those bones in mammals too (the third bone of the inner ear, the stapes, is sourced from the second pharyngeal arch of the human foetus). Not just your jaw, but your ear too, then, share their origins with the gill slits of those ancient Chengjiang *Myllokunmingia*.

Jaws turned out to be more widespread in those early days. Appearing a little later than the placoderms (and probably descended from them) were the 'spiny sharks', or acanthodians. These did not yet compete with *Carcharodon* in size or grandeur—they were normally just a few centimetres long—and the name 'spiny' comes from the stout spines that acted as fin supports and which turn up, for instance, as the most common 'large' fossil fish fragment (i.e. up to a centimetre long—but still regarded as a very nice find by collectors!) within the celebrated Ludlow Bone Bed of late Silurian age. Nevertheless, like the modern sharks they probably gave rise to, they had a skeleton of cartilage, and jaws with teeth (real teeth, unlike the sharply sculpted jaws of the placoderms), and had streamlined shapes for active swimming. They were mostly mini-predators rather than super-predators perhaps, yet nevertheless they enjoyed more than 100 million years of success.

Jaws characterize us, and other mammals, reptiles, and amphibians, and the bony fish that dominate the marine and freshwater realms today. Nevertheless, it used to be thought that the placoderm and acanthodian jaws were a dead end, which went extinct with these groups of fish in the Palaeozoic Era, and that the jaw design independently arose in the bony fish: the respective jaw structures were thought too different in detail for one to have evolved directly from another. But, in 2013, Min Zhou and his colleagues found a finely preserved, almost uncrushed late Silurian placoderm fish in China that they called *Entelognathus primordialis*—a beautifully appropriate name meaning 'primordial complete jaw'. This clearly showed equivalents of bones—like the premaxilla, maxilla, and dentary bones—possessed not just by bony fish but the limbed vertebrates, too. This is a very recent discovery and is still provoking debate. But, the origins of our own jaws, on current evidence, may well stretch back to the extraordinary placoderm fish that so bewitched Agassiz.

The placoderm armour, too, showed innovation. Unlike the enamel- and dentine-like armour of the jawless ostracoderms, it was largely made of cellular bone. As a living skeleton, therefore, like our own, it could be remodelled during growth. There were the beginnings, too, of a mineral- ized internal skeleton. The placoderms, like the first sharks, possessed true vertebrae. These vertebrae were mostly of cartilage but, unlike those of sharks, they had mineralized, bony arch-like skeletons. Placoderms also had two sets of paired fins, pectoral in front and pelvic behind. This seems to represent the origins of the fins of the bony fish—and of our arms and legs, too.

Then, there are the bony fish that are dominant today. The earliest bony fish appear in rocks of about the same age that *Entelognathus* was excavated from. *Guiyu oneiros*—the first part of its names means 'ghost fish'—is more of a chimaera than a ghost (Figure 18). It possesses characteristics of both of the two groups of living bony fish, those that are ray-finned and those that are lobe-finned, but it has more of the latter characters than the former. Ray-finned fish have their fins supported by bony spines that connect to the skeleton below. They comprise more

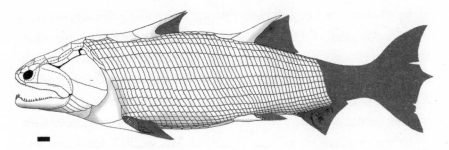

Figure 18. The early fish *Guiyu oneiros*. Scale bar is 1cm.

than 99% of the diversity of modern fish, with something like 30 000 species—making them by far the most numerous vertebrates. But whilst the skeletons of ray-fins have adapted to a huge variety of forms, from a tiny minnow to the gigantic 4-metre-long (and 4-metre-tall) marine sunfish, it is the lobe-fins that have literally inherited the Earth. Within the structure of their skeletons is the foundation of the four-limbed animals that would one day walk on land.

Half Fish, Half Tetrapod

If lobe-finned fish were the antecedents of all vertebrates that walk on land, then palaeontologists would predict that somewhere in the fossil record we should be able to find the skeleton of an intermediate between a fish and something looking like an amphibian. Such a fossil animal would be a real chimaera, a kind of half fish and half tetrapod, or as some have called it, a 'fishopod'.

The lobe-finned fish, known as fossils from late Silurian times, used to be thought absent from modern oceans—long absent, indeed, for more than 60 million years—until, famously, their representatives, the coelacanths, were rediscovered living in the depths of the Indian Ocean in 1938. The lobe-fins include the lungfish too, which today all live in freshwater. The few living lungfish species are widely dispersed between the continents of South America, Africa, and Australia, the remnants of a much wider diversity of lungfish that characterized the ancient continent of Pangaea. They are also the last survivors of one of the great colonization events in history, the conquest of land by the vertebrates. Lungfish belong to the group of animals from which all four-limbed vertebrates, humans included, evolved.

Lobe-finned fish are characterized by their well-developed fleshy paired pectoral and pelvic fins. These four fins, probably an inheritance from the placoderms, are the framework from which four-legged land animals evolved. These muscular fins articulate with the main skeleton of the fish via a single bone. This differs markedly from the ray-finned fishes, where the fins have several bones at their bases. Nevertheless, modern lobe-finned fish may not be the best analogues to indicate the origins of four-limbed terrestrial animals. For that, we need to find some ancient fossil lobe-finned fish. In fact, we need to find that 'half fish–half tetrapod' ancestor, the 'fishopod'. And just such a fossil is Tiktaalik.

Tiktaalik (the name was suggested by local Inuit elders as their name for a kind of freshwater fish) is a fossil from the Arctic northeast of Canada, from the icy wastes of Ellesmere Island. Looking superficially like an alligator (Figure 19), and up to 3 metres long, Tiktaalik is nevertheless classified as a lobe-finned fish. It lived along ancient tropical rivers some 375 million years ago, at a time when vegetation was spreading widely across the landscape. Tiktaalik had both gills and lungs. It had fish features such as scales and fins, but a flat head that looks more like an alligator or crocodile. More importantly, its skeleton shows the early development of a mobile neck, of ribs to support its body, and of a wrist joint: all fundamental to the development of a skeleton that can take vertebrates on to land. Tiktaalik may have spent most of its time in shallow water, perhaps briefly emerging to drag itself across the ground above the waterline, much like a mudskipper today.

Tiktaalik is an animal predicted to have existed by the scientists involved—and then, triumphantly, found.[33] But it may not be our direct ancestor, hence not a 'missing link'. A few years later, footprints were recovered from

Figure 19. Reconstruction of *Tiktaalik*. The animal was up to 3-metres long.

a quarry in Poland that suggests some tetrapod (its bones have not yet been found) was walking on—not dragging itself along—land nearly 20 million years earlier in the Devonian Period. The mid to late Devonian was clearly a time of evolutionary ferment around the boundary of land and sea. The late Devonian saw marked environmental change, when the oceans were undergoing long periods of poor oxygen levels. Animals in shallow water may have been more secure from such physiological stress. Those with the right characteristics to adapt to straying onto land would over time modify fins to eventually operate as limbs, lungs to breathe air—an ancient trait of fish—and the mobile head that is characteristic of tetrapods.

By 365 million years ago, at the very end of the Devonian Period, there are more 'fishopod' animals to add to this story, though by now these were a little less fishy, with significant developments around the 'pod' end of things. This is the time of animals such as *Ichthyostega* and *Acanthostega*, flat-skulled animals up to 2 metres long. Their appendages look more limb like than those of *Tiktaalik*: no longer just large muscular fins, these had distinct separate digits, albeit in greater number than we are used to (with seven in *Ichthyostega* and eight in *Acanthostega*). These animals still did not stand up on their legs like a true quadruped (which makes the Polish footprints all the more mysterious) but could crawl across the ground, and they stayed in or close to water.

For the first animals that could move their limbs like a quadruped, one has to travel forwards in time again, to the Carboniferous Period that began about 360 million years ago. This is a time called Romer's Gap, a supposed 15-million-year gap in the fossil record at the end of the Devonian and beginning of the Carboniferous where there is a paucity of fossils tracking evolution from the lobe-finned fish and crawling tetrapods of the Late Devonian, to the more advanced tetrapods of the Carboniferous. This 'gap' is named for the American palaeontologist Alfred Romer, who was the first to spot it.

This gap has now been filled by ancient skeletons discovered by the amateur fossil hunter Stan Wood, working with Cambridge scientist Jenny Clack. They have unearthed and studied many new fossil tetrapods in 350-million-year-old sedimentary rocks of southern Scotland. These fossils include the tiny *Casineria*.[34]

Casineria is an animal of great anatomical significance to the human hand, being the earliest animal that possessed a forelimb terminating in five digits. This, along with other skeletal features of *Casineria*, suggests that this animal truly walked on land. From here on, with the first fully land-based tetrapods, it is but a few small anatomical steps to the 30-metre-long titanosaurs of the Cretaceous Period. But before then, vertebrates had to shift their legs beneath the mass of their bodies and develop means of reproducing successfully away from water. They also had to survive the environmental calamities of the Permian–Triassic and Triassic–Jurassic boundaries, some 100 and 150 million years after *Casineria* walked the land. These and the later extinction of the Cretaceous–Paleogene boundary would selectively promote different groups of tetrapods, the archosaurs at the Permian–Triassic boundary, the dinosaurs at the Triassic–Jurassic boundary, and then finally the mammals at the Cretaceous–Paleogene boundary some 66 million years ago.

Before all of this, there is still the skeleton—that we call the shell—of an egg to consider.

Amphibians to Amniotes

Modern amphibians, and their immediate tetrapod ancestors needed water to reproduce, even if they were making their first steps onto land during the Carboniferous. This has always limited their distribution, though this has not stopped frogs and toads from becoming a geologic- ally long-lived and diverse group, with something approaching 5000 living species. But for animals to become fully terrestrial, and to move away from laying eggs in water, a watertight egg that can be deposited on land was needed. Such animals are the amniotes, the group that includes reptiles, birds, and mammals. Arthropods had evolved their own version of eggs long before, with examples of egg-laying evident in some of the Cambrian animals from the Chengjiang biota. That egg- laying capacity probably gave arthropods an inbuilt advantage when they began to colonize the land during the Silurian. For vertebrates though, the very invention of the amniotic egg itself is directly linked to the colonization of land.

It is the amniotic egg, and the lack of a larval stage—the tadpole— prior to metamorphosis to the adult that clearly sets amniotes apart from amphibians like frogs and newts. Sealed on the outside by a leathery or calcified skin, the amniotic egg is a marvel of engineering, within which the embryo is protected by a series of membranes, one of which—the chorion—allows for the gaseous exchange of carbon dioxide that would otherwise suffocate the developing animal.

The tiny *Casineria*, at less than 10 centimetres long as an adult, may have been amongst the first of the amniotes. Its very smallness is crucial, for the first amniotic eggs deposited on land were probably very simple and small. They would have evolved from the eggs of amphibians, and though they had formed a leathery outer skin to protect against desiccation, they probably lacked the functionality of later vertebrate eggs that allowed gases to diffuse via the chorion. This inability to actively dispose of a gas like carbon dioxide would have limited the egg size to less than 1 centimetre in diameter—a size that allowed carbon dioxide to diffuse out of the shell naturally. A small egg means a small full-sized animal, and so such eggs did not produce the metre-scale tetrapods of the latest Devonian that still laid their eggs in water. Tiny *Casineria* then, may be another important staging post in the conquest of land. It was not alone, as there are other tiny contenders as amniotes amongst the tetrapod fauna of the early Carboniferous. To evolve from tiny *Casineria* to giant reptiles, though, was to require the evolution of much larger eggs.

Very early in the evolution of amniotes, perhaps as much as 320 million years ago, two major groups appeared, the sauropsids and the synapsids. Both of these groups were to have a profound influence on the terrestrial biosphere, and in turn some of their descendants were to go on to colonize the oceans, in most spectacular fashion. One of those groups, the sauropsids, has the telltale signature of 'dinosaurs' hidden within its name. The other group, the synapsids, would go on to give rise to an array of animals that includes blue whales, mammoths—and humans.

The 'Not Quite Dinosaurs' of the Permian

It is often said that for mammals to flourish, they had to wait for the non-avian dinosaurs to go extinct in the end-Cretaceous apocalypse. Much earlier, though, during the Permian Period, it was the ancestors of the mammals, the earliest synapsids, that were the dominant carnivores, not the sauropsids.

The plastic dinosaur kits that appeared in the Christmas parcels of 1970s 10 year olds typically featured a *Tyrannosaurus*, a *Brontosaurus*,

Figure 20. *Dimetrodon.*

a *Stegosaurus*, and also that distinctive sail-backed beast *Dimetrodon* (Figure 20). As a child of the 1970s, one did not discriminate *Dimetrodon* from the dinosaurs; after all, it had done battle with a suspiciously lizard-like dinosaur in the original 1940 movie *One Million Years* BC,[35] which at the age of 10 years, amounts to wholly convincing evidence of taxonomic affinity.

Dimetrodon, though, lived about 290 million years ago, long before the dinosaurs. And, though it possessed a lizard-like body with that characteristic sail running along the length of its back, and a fearsome array of sharp teeth, it is in fact more closely related to the mammals than it is to the dinosaurs. The tell-tale signs are in its skeleton, with a skull that develops a hole, the 'temporal fenestra' behind each eye socket, and also in its mouth, where the teeth are differentiated as canines and incisors. These are features that are also possessed by mammals.

The famous sail remains enigmatic. Skeletally, it is made up of a series of long spines growing upwards, one from the top of each vertebra. In life, it is assumed that these spines supported a web of tissue, to make the sail. Proposed explanations have included that it really was a sail, for use when the animal was swimming, or that it helped camouflage the animal among reeds, or it was a stabilizing device as the animal walked, or that it was a heat regulation device, to absorb sunlight in the morning and release heat in the evening. That last explanation has long seemed most plausible—but modelling the thermal regime of *Dimetrodon* suggests that the sail would not have been a very efficient regulator. Rather, it has been suggested, the sail may have been used to attract the opposite sex—and so perhaps was an early, and bony, forerunner of the peacock's tail.

Dimetrodon and its sail-bearing kin are not the direct ancestor of the mammals, but like *Tiktaalik* this was another staging post, in this case along a road that would eventually lead to mammals. In due course more advanced synapsids would evolve, called the therapsids. These animals are notable for skeletons where the limbs are positioned beneath their bodies, unlike the sprawling lizard-like limbs of *Dimetrodon*; synapsid feet were oriented parallel to the animal's axis too, rather than splaying outwards in lizard-like fashion. Amongst these therapsids are the cynodonts, the group from which modern mammals evolved.

The large, successful land predators of the Permian Period were, then, the synapsids. It was only in the subsequent Triassic Period that sauropsids triumphantly gave rise to the myriad dinosaurs and crocodilians that dominated Mesozoic Era landscapes. Their rise took place in the aftermath of calamity.

Enter the Dinosaurs

Two hundred and fifty-two million years ago, and perhaps within a brief interval of a few tens of thousands of years, animal Earth nearly died. The precise interplay of causes of this mass extinction is still hotly debated, but trigger factors included a toxic mix of volcanic eruptions on a continental scale covering much of modern Siberia, the acid rain that ensued from this, extreme ocean anoxia, and a hyper-greenhouse climate. Palaeontologists calculate that more than 95% of the world's marine species went extinct, and perhaps 70% of terrestrial vertebrate species too.

As the world began its recovery in early Triassic times, one previously obscure group of the sauropsids, the archosaurs (meaning 'ruling reptiles'), replaced the devastated synapsids as the dominant land vertebrates. Archosaurs possessed skulls with two openings ('temporal fenestrae') behind the eyes, thus distinguishing them from the synapsids, which have one. This double-hole arrangement allows for stronger attachment of muscles to the jaw, allowing the mouth to open wider. Though these holes have been lost in descendants such as snakes and birds, they enabled development of the magnificent jaws of a *Tyrannosaurus rex*, and of the fearsome gape of living crocodilians.

From root archosaur stock, two major groups were to evolve during the Triassic: that which includes the modern crocodiles and many extinct groups, and that which gave rise to the dinosaurs, including birds. Looking into a crocodile's mouth, and then staring into the toothless beak of a sparrow, one might think that these two are unlikely relatives. But they have a shared history going back some 250 million years, and within such a long span there was enough time for many different skeleton types to evolve. Of these two groups, it was not the dinosaurs that first wrested supremacy from the Permian synapsids, but the crocodilian branch. It took another mass extinction event at the Triassic–Jurassic boundary 201 million years ago—again associated with a brief but extraordinarily intense phase of volcanicity—to partly collapse the Earth's biosphere once more and, following this, the dinosaur branch rose to supremacy. For the next 135 million years, dinosaurs would become the giant walking skeletons of Jurassic and Cretaceous landscapes.

There are something like 10 000 dinosaur species living now, which we know as birds, making them the most diverse living tetrapod group. Estimates of the number of dinosaurs that roamed the Earth in the past are more difficult, not least because of the incompleteness of the fossil record. But a recent scientific study[36] estimated that a little under 2000 dinosaur species inhabited the Earth during the Mesozoic Era. This is a surprisingly low estimate given that an average species span might be something on the order of 5 million years, and so this suggests that at any one time only something on the order of 75 species of dinosaurs coexisted on the Earth.

These dinosaurs showed great variation in skeletal design along two broad themes: the saurischian 'lizard-hipped' branch, which included the therapods and sauropods, and the ornithischians, or 'bird-hipped' forms, which include *Triceratops* and *Stegosaurus*.

A Question of Size

Dinosaurs have such a tight grip on human imagination not just because of the bizarre forms they took, but because of their size—some were genuinely, outrageously, almost unbelievably enormous. The greatest of all were the sauropod titanosaurs, which reached over 30 metres in length, with skeletons supporting animals that weighed perhaps 90 tonnes, or the equivalent of 15 African elephants in one animal. The huge size was partly a defence against predation, although the predators did their best to keep up, and included such as *Tyrannosaurus* and *Gigantosaurus*, adults of which could grow more than 12 metres long, nearly 4 metres tall, and weigh over 10 tonnes. How was an animal able to grow so big? Much of the answer resides in the skeleton.

First of all, the bioengineering constraints imposed by adopting a life on land shaped and optimized the fundamental framework. For any sizeable adult just to be able to stand up and walk or run in air, rather than be cradled in water, required the succession of skeletal readjustments and reinforcements that took place from *Tiktaalik* onwards.

To go from functional at what might be regarded as normal size, to reach the extremes of the sauropod dinosaurs, further measures were needed. To support a massive sauropod skeleton requires tree trunk-like limbs, with broad hind feet, and horseshoe-shaped front feet. In many species the vertebral bones, including those of the neck, had air pockets to make them lighter—and to help carry the air the 10 metres or more from the mouth to the lungs. And to develop their huge necks further they reduced the weight of their skulls, so that these became little more than food-processing mechanisms. It has been suggested that some sauropods adapted this skeleton to allow them to rear up on their hind legs to find food, using the long tail and limbs to form a stable tripod. Sauropods have been given names to match their fantastical bodies, such as the African *Giraffatitan*, the 'giant giraffe' that stood some 12 metres tall.

The strategy of building gigantic skeletons clearly worked for the sauropods, for many millions of years. Though they would have been slow-moving animals, their massive size helped protect them from attack. They did not survive the end-Cretaceous mass extinction, so only their titanic skeletons survive now, to fill the cathedral-like galleries of our major museums, there to be stared upon with awe by animals possessing considerably more modest skeletons.

Reclaiming the Seas

Life arose in the seas, almost certainly. The Earth's biosphere had been almost exclusively marine for over 3 billion years before it extended out over land. The first vertebrates arose in the sea. And yet, when it comes to hogging the top of the food chain in the seas, or to breaking records for size and weight, then it helps to have an ancestor that has served time on land. That was true in the Mesozoic with the marine reptiles, and it is true today with the cetaceans—or at least it was until recent human interference. What is it about the terrestrial skeleton that makes the sea an easier place to conquer?

Louis Agassiz had kind words for Hugh Miller, the self-educated stonemason. He was equally generous to—and a touch astonished by—a pair of even more unlikely and precocious British 'fossilists', Elizabeth Philpott and Mary Anning, of Lyme Regis in Dorset. Visiting

Lyme Regis in 1834, he was on the trail of fossil fish. Anning then was in her mid-30s and already established, or perhaps notorious, as the local excavator, and restorer—and interpreter too—of the fossil skeletons of ichthyosaurs, plesiosaurs, and pterosaurs from those crumbling Jurassic cliffs, as well as of the small fry, the ammonites and belemnites that she sold for a few shillings to keep herself and her family—just—out of abject poverty. Philpott, nearly 20 years older and from a rung or two higher in society, was the friend and mentor to the untutored young woman from a 'dirt-poor' background, and she specialized in the fossil fish. When Agassiz visited, the two women could show him 34 distinct species of fossil fish—and discuss the identification of their skeletal fragments 'with utter certainty', as the Swiss-born savant later recalled.

The Lyme Regis fish were beautiful, lustrous fossils. But public and scientific fascination inevitably focused on the monstrous reptiles that Mary Anning dug out of the cliffs. In 1830, the eminent geologist Henry De La Beche—who was soon to become the founding Director of the Geological Survey of Great Britain—painted the first widely known diorama of ancient life, *Duria Antiquior* ('Ancient Dorset'). In this, Jurassic fish were depicted swimming through Jurassic waters, and ammonites and belemnites too, but it is the larger beasts that grip the eye, from the ichthyosaur seizing the neck of an unfortunate plesiosaur, to the crocodiles and turtles in the shallows, to the pterosaurs flying overhead (Figure 21). The painting was no mere academic exercise, for De la Beche and Anning had history. As a boy, De La Beche lived in Lyme Regis, befriended the young, fossil-collecting girl, and, it seems, gained his love of geology from her. Now, he was repaying the favour: prints of *Duria Antiquior* were sold to provide funds for the Anning family, who were then, as so often, in financial difficulties.

Duria Antiquior may be a first sketch of a past world, with an eye to drama and to the paying public. But its message is a reasonable representation of Jurassic reality: it was not fish that were the apex predators in the seas, but formerly land-based tetrapods, returning to the realm of their very distant ancestors. And this was no isolated freak interval in

Figure 21. *Duria Antiquior.*

geological time. With brief interludes (and admittedly with some stiff competition, such as from the sharks) this has been the case for the last 250 million years.

It has been quite a procession. Over this quarter of a billion years, some 30 groups of land-living tetrapods have successfully made a transition back to the sea. Early in the Triassic, the ichthyosaurs and plesiosaurs, together with another group, the thalattosaurs, grabbed the marine ecospace left wide open by the end-Permian catastrophe. In the Jurassic, the turtles and the crocodiles joined the invasion. These groups held their dominance for a long time—the plesiosaurs right through the Mesozoic Era, while the ichthyosaurs only declined in the Cretaceous Period, when the mosasaurs appeared, to outmuscle and replace even those fierce reptiles. And subsequently, in the Cenozoic Era that we still live in, there have been waves of invasion from the seals, penguins, cetaceans, and others.

It was not just a procession, but one-way traffic. No major terrestrial vertebrate groups were to emerge from the sea over that time, to follow in the footsteps of *Tiktaalik* and its kin, to take over the land from current incumbents. And the invasion from land to sea clearly took place despite considerable hurdles: the body and skeleton had to be remodelled to no longer deal with the pull of gravity in air, but to enable quick and effective movement through that denser fluid, water. There was the question of respiration, too: none of the invaders re-evolved gills, and so had, and still have, to come to the water's surface to breathe.

That, though, brings us to one of those exceedingly difficult 'why' questions in science—why did the tetrapod invaders, initially ill adapted to the water, successively establish dominance over the organisms in which the whole ancestral line was devoted to the honing of perfect adaptation to life in water?

As far as we know, there is no straightforward answer to this simple question. The high metabolic rates of mostly warm-blooded animals may have been one factor. The tetrapod skeleton, evolved, refined, and tempered to deal with conditions on land, probably also had something to do with it.

That skeleton, for whatever reason, seemed to have been profoundly adaptable. Early in the Triassic period, as the early ichthyosaurs, plesiosaurs, and thalattosaurs established themselves in the marine realm, a variety of skull types emerged among them, from large-jawed macro-predatory forms, to those adapted to shell-crushing, to fanged jaws for fish-catching, and toothless jaws for suction feeding. The bones as a whole adapted, becoming denser in those forms that stayed close to shore, to allow for buoyancy control, and less dense in those that ventured far into ocean waters, to minimize energy loss in swimming. Limbs evolved back into paddles or fins, while the body shape became more streamlined, with the ichthyosaurs in particular assuming a classic 'dolphin' shape (although it is really the dolphins that are now assuming a classic ichthyosaur shape).

And as for size, the marine reptiles took the palm too. They had one close contender: the enormous Jurassic fish *Leedsichthys*, named not after the Yorkshire city, but after the 19th-century collector Alfred Leeds, who uncovered much important material of this form. *Leedsichthys* perhaps reached some 17 metres long—though it was, temperamentally, a pussycat, being a placid filter-feeder, much like the whale shark today. The largest mosasaurs reached a similar length, but were emphatic super-predators, while the ichthyosaur *Shastasaurus*, the largest marine reptile known, reached 21 metres,

It was quite an empire. But it fell apart, 66 million years ago, as a giant meteorite struck what is now Mexico, precipitating a calamitous mass extinction, with the loss of perhaps 75% of all Earth's species. While survivors among land vertebrates included the turtles, crocodiles, and birds, new empires were to be built on the ashes of the old one. One group of tetrapods had played the long game. It was about to re-inherit the Earth.

The Making of the Modern Skeleton

After the Mexican impact 66 million years ago, the Earth's biosphere was yet again collapsed—it was the fifth such major reshaping in half a billion years—with the loss of many of its main components of Mesozoic times. Among the vertebrates, the non-avian dinosaurs perished—but their avian kin, the birds, were set to flourish, and to recolonize the land as, among a large variety of forms, the giant 'terror birds' of South America. Also surviving into this landscape, though not yet standing tall, were the mammals. One of these groups was the primates—ultimately to evolve into tarsiers, lemurs, monkeys, apes, and humans—the last member of which was to have—and continues to have—the most profound impact on Earth. The mammals were now set to steal the stage from the dinosaurs, but their origins lie much earlier, before the dinosaurs appeared, in the form of a group of dog-like carnivores in the Late Permian called the cynodonts. They possessed a small but useful skeletal innovation—that we still make use of every day.

There is evidence in the skulls of cynodonts for the development of a secondary palate. This is a structure that, in humans, develops early in pregnancy at about six weeks. It is the structure that forms the roof of the mouth, separating airflow there from that travelling through your nose—it means that you can breathe and eat at the same time. Possession of a secondary palate is a primary feature of mammals. But cynodonts still had some characteristics of their earlier ancestry too, because they continued to lay eggs, a style of reproduction that is echoed in their distant relatives, the living monotreme mammals of Australasia, which include the duck-billed platypus and the spiny anteaters or echidnas.

The cynodonts survived the end-Permian calamity. But their path towards the evolution of mammals now unfolded amongst the Triassic survivors of a post-apocalyptic landscape. Early Triassic cynodonts already possessed some of the behavioural and social skills that would set them out as survivors. The small cynodont *Trirachodon* of southern Africa lived in social groups in complex burrow systems.[37] Like prairie dogs of the Triassic, these animals had social skills that may have supported protection from predators, the rearing of young, and perhaps even avoidance of marked temperature changes at the surface between day and night, or between the seasons.

But, in tracing the ancestry of mammals from the cynodont survivors of the Early Triassic, few of the characteristic features we might regard as mammal, like mammary glands, glands for grooming hair, or intellectual smartness, are easily preserved in the fossil record. The skeleton can readily be fossilized, though, and the bones of the jaw and middle ear now become decisive in tracing mammal origins.

Mammals use two bones for hearing that all other amniotes use for breathing. In mammals the jaw joint is composed only of a lower dentary bone and this articulates with the squamosal bone in the skull. But in reptiles there are intervening bones between the dentary and squamosal, called the quadrate, articular, and angular bones. In mammals, small vestiges of the quadrate and articular bones still exist, but they form the incus and malleus bones of the middle ear. The third bone of the inner ear of mammals, the stapes (or 'stirrup')—the smallest bone in the human body—is the same as the columella bone of all reptiles. These structures are 'half-way' formed in the skeleton of the mouse-like Late Triassic *Morganucodon*, which lived in South Wales more than 200 million years ago, but they are found for the first time fully formed in the tiny, shrew-like Early Jurassic mammal *Hadrocodium* that lived just a little later, some 195 million years ago.

This small but crucial skeletal structure links all three living groups of mammals: the monotreme echidnas and platypus of Australasia and New Guinea that lay eggs; marsupials such as the kangaroo, which complete most of the gestation of young within an exterior pouch; and the placentals, such as cows and sheep, which give birth to well-developed young that in some cases are able to walk within hours.

A New Skeleton Diversity

With the dinosaurs out of the way, the mammals soon took over apex positions in the Earth's biosphere, both land and sea. With their ability to regulate their own body temperature, their intelligence, and their marked ability to cooperate in groups, they were well placed for this position. Their initial small size was probably a factor aiding survival in the meteorite impact's aftermath, but greater physical dimensions were to come, most emphatically. On land, there developed the evolutionary lineages of animals both familiar—the horses and elephants, cats and dogs—and less familiar—such as the brontotheres, titanotheres, and uintatheres. These now provide some of the textbook examples of evolution. Invasions of the sea continued and intensified—with the seals and walruses, penguins, whales and dolphins, manatees, and others; and, most recently—beginning just some 6 or 7 million years ago—the sea snakes. Marine tetrapods climbed in species richness through the Cenozoic, even farther above the levels reached in the Mesozoic.

The marine invasions continued despite—or perhaps in part because of?—further changes to the skeleton of the bony fish. These now became the familiar highly mobile, lightly armoured forms we are familiar with, and thrived in the Cenozoic oceans, lakes, and rivers. Unlike us, they still retain extensive remains of the 'tooth-structured' external coverings of the ostracoderms that Louis Agassiz and Hugh Miller puzzled over, though these are now mostly reduced to arrays of delicate scales, perhaps in part as an adaptation for speed. In essence, the modern vertebrate world was here. Late in the Cenozoic, there was further innovation. To paraphrase George Orwell: four legs good, two legs (in some significant ways) better.

All Fours to Upright

There are currently (in the year 2017) about 7.35 billion humans on Earth. Their distribution extends from the tropics to the polar scientific bases of the Arctic and Antarctic. They live on the sea, beneath the sea as submariners, and at all altitudes to 5100 metres, the height at which the people of La Rinconada in the high Peruvian Andes live. If you take the average human weight at about 62 kilograms (it rises to over 80 kilograms for people from some of the wealthiest nations), then massed together, the weight of humanity currently measures over

450 million tonnes. That is about one-third of the total mass of all land vertebrates, with the great majority of the other two-thirds being taken by the animals—cows, pigs, sheep, goats, chickens—that we keep alive (briefly) to eat. Wild animals make up some 5% or less of that global vertebrate biomass. What component of that mass is skeletal? The average human skeleton weighs about 15% of the total body's mass. That means that human skeletons currently weigh in at about 72 million tonnes, with perhaps another 150 million tonnes of the bones of our captive prey animals.

That is now an unnaturally large amount of bone. The American palaeontologist Tony Barnosky has estimated that total vertebrate biomass on land has gone up by about an order of magnitude over a natural baseline. This massive growth has been unnaturally fuelled by our species' own ingenuity not least. Humans have had to grow food plants more prolifically, to feed more efficiently to animals, to then feed to us. Many components are involved, but we might focus briefly on that key ingredient of bone, phosphorus.

In 19th-century Britain, farmers, struggling to feed a population that doubled in that turbulent century, noticed that crops grew better near graveyards. It was just a short step to adding ground-up bone from the slaughterhouses and knackers' yards to the fields. More was needed. It was yet another short step to grinding up human bone, taken from the remains of hussars, grenadiers, dragoons, scavenged from the battlefields of Europe. At its peak the trade took about 3 million cadavers a year across the Channel, to feed the hungry British. The great German chemist Justus Liebig was horrified, and said Britain was like a vampire, hanging on the neck of Europe. It was not just human corpses that were so recycled. A store of 180 000 mummified cats was found in Egypt, dating back to the times of the Pharaohs; shipped to Liverpool in 1890, they were also ground up to put on to English fields; the magazine *Punch* published a cartoon of spectral bandage-bound cats glaring balefully on terrified farmers, as vengeance for such maltreatment of their earthly remains. More ancient bones came into play then, too, for the farmers. Snaking beneath the soils of eastern England, there are thin strata rich in the fossilized bones and teeth and faeces of marine reptiles and dinosaurs—that became a thriving trade too, and a palaeontologist might weep to think of so many superb specimens crushed to help the growth of turnips. We are still digging rock phosphate out of the ground to grow our own bones. It is a transgenerational and transepochal transfer of singular planetary scale—and we still depend on it utterly. Scientists have lately written on an impending 'peak phosphorus', as the capital seems to be running out uncomfortably fast.

The Roots of Human Abundance

How did we get to the point where one type of animal skeleton, the human one, accumulated so much mass?

Humans are great apes, nestled within that wider grouping of mammals called the primates that originated in the Paleogene Epoch, some 60 million years ago. Living primates are a diverse group that is, with the exception of humans, distributed through the tropics and subtropics. After humans, the next most populous group of primates is Mueller's Bornean Gibbon, with between 300 000 and 400 000 individuals, enough to populate one moderate-sized human city like Bristol, England. This difference from the 7-plus billion humans is immense, and as we move down the list of primate species abundances, the contrast will become yet starker. Abundance for chimpanzees is a population close to that of the English city of Plymouth (about 200 000–300 000 individuals), the port from where the Pilgrim Fathers sailed. The 'man of the forest', the Bornean orangutan, may account for some 70 000 individuals, or the population of the English seaside town Torquay. The Sumatran orangutan fares less well, for Sumatra is less remote from human influence than Borneo, and the pressure for space and forest is greater. Here, there are only about 7000 orangutans, the population of a couple of villages. Down the list of primates, the numbers become diminishingly small, with some gibbon species contributing a hamlet's worth of inhabitants (perhaps just 20 or so individuals), or worse, the inhabitants (family and servants) of a single wealthy human home. One primate species, humans, constitutes more than 99.9% of all primate individuals on the planet.

This explosion of humans is new. There were about 1 billion humans in 1800, and perhaps just 200–300 million of us at the time of the Romans. Go back further, to the super-eruption of the gigantic Sumatran volcano of Toba some 70 000 years ago, and perhaps only a few thousand modern humans survived in the horn of Africa. Even then, that is late in the history of the human genus *Homo*, a story whose roots developed over more than 6 million years. In the early days too, there was a much greater variety of early humans and human relatives, though these species are now found only as fossils. Our last close relative, the Neanderthals, disappeared from western Europe some 40 000 years ago.

The human skeleton is a modification of the tetrapod animal that crawled out of the sea some 350 million years ago. In that design was the wellspring of intellectual advance—a hand with five fingers. But first, our ancestors needed to walk upright. That was to free the forelimb to interact with eye and brain.

All great apes, bar humans, spend most of their time walking on all fours. It is the characteristic walk of mammals and of tetrapods. Chimpanzees and gorillas, for example, are often referred to as knuckle walkers. Indeed, only a relatively small percentage of the many thousand fossil and living tetrapods seem to have mastered walking on their hind legs, though the ability to do this seems to be deep-rooted in the genetic codes that control the formation of our skeleton. The oldest example of this is in the Permian reptile *Eudibamus*, which lived in ancient Germany some 280 million years ago. Small—25 centimetres long—but perfectly formed, its strong hind limbs and small forelimbs suggest its bipedal gait.

For primates though, to consistently walk on two legs, and to do this in a fully vertical position, is the domain of humans of the genus *Homo*, of their near relatives, the australopithecines, and of some of our great ape cousins that began to walk upright perhaps 6 million years ago. Where and why did this skeletal change begin?

The early story of our ancestry is found in a mosaic of fossils from Africa, from between about 7 and 3 million years ago. One part of this mosaic is *Sahelanthropus* (Figure 22), which lived in West Africa some 7 to 6 million years ago.[38] It is known from only a few fossil bones, but importantly these include a skull and a mandible: it is possible, therefore, to look into its tall face with its massive brow ridge and sense the human story beginning. Looking at the base of the skull, where it connects with the spine, one can make more specific deductions.

Figure 22. The cranium of *Sahelanthropus*.

This connection is forwardly positioned. It suggests that *Sahelanthropus* stood upright on two legs.

The known mosaic of early human relatives in Africa is now extensive, the most famous of these early human relatives being the Australopithecines. These appeared in East Africa about 4 million years ago, became highly successful, and survived to overlap in time with the earliest humans. Australopithecines and their kin, *Kenyanthropus*, seem to be the first of our relatives to have used stone tools as extensions of their modest skeletons, about 3.4 million years ago. And it was probably Australopithecines that made the footprints preserved in the 3.7-million-year-old volcanic ash at Laetoli in Tanzania, which track three upright individuals—perhaps a single family—among tracks of many other animals, including hyenas, rhinos, and baboons. Our early human relatives, small and weakly equipped anatomically, were wandering in a landscape of megafaunal diversity.

Into this late Pliocene landscape walked *Homo*. The oldest fossils of our direct ancestry come from near the crest of a flat-topped hill in the Lee Adoyta region of Afar, Ethiopia.[39] The fossil is a jaw with teeth, and is 2.8 million years old. Two other species of *Homo*, *H. habilis*—the handy man—and *H. rudolfensis*, turned up in strata about 2 million years old in East Africa. A little while later, about 1.8 million years ago, came *Homo erectus*. These newly evolved humans had larger braincases, smaller jaws and teeth, and shorter limbs than the Australopithecines. Humans had now well and truly arrived in the landscape.

From Hand to Mouth

Standing upright was fundamental to the development of human intelligence. It freed the hand to develop opposable thumbs, and so the human skeleton can grip and manipulate tools, and undertake complex and precise procedures such as brain surgery, or turning the pages of this book.

The third metacarpal styloid process is the key to this.[40] It is a structure that sits at the base of your middle finger, where it helps the hand to lock into the wrist. It confers strength to the hand and allows it to manipulate stone tools—or a power drill—with great dexterity. This third metacarpal process is absent from the hands of all non-human primates, and from the earliest humans too: *Homo habilis* may have been 'handy', as the name suggests, but—not possessing the third metacarpal styloid process—only up to a point.

Evidence for the third metacarpal styloid process is found in *Homo erectus* fossils about 1.4 million years old from the classic fossil site of Kaito, in West Turkana, Kenya. This skeletal innovation is probably linked to new kinds of stone tool making, notably the Acheulian culture, a successful technology that first appears in the sedimentary strata of East Africa about 1.75 million years ago, and survived for another 1.5 million years. Acheulian tools are more sophisticated than earlier tools, and have shaped bifacial facets that required both dexterity and imagination to manufacture: in this process, hand, eye, and mind were linked.

During the period from about 1 million years ago to 200 000 years ago nature experimented with several different skeletal designs for humans. In Europe, this included the Neanderthals, whilst in parts of East Asia *Homo erectus* persisted until recently. Then, about 300 000 years ago a new species emerged in Africa, *Homo sapiens*.

Anatomically modern humans appear first in the fossil record of Morocco. However, although these early humans possessed the same overall skeletal structure as us, they continued to evolve, so that scientists differentiate 'anatomically modern' from 'behaviourally modern' humans that appeared later, about 70 000 years ago. What then was happening inside the head of these early *Homo sapiens*?

The human skeleton has remained essentially the same for 300 000 years, but other aspects of the human condition were clearly changing, as is evident from the evolving stone tool kits of the Mid and Late Pleistocene, and which therefore probably involved feedbacks between the hand and brain. By 70 000 years ago there is evidence that humans had developed symbolic behaviour, clear from the fragments of ochre and bone on the floor of the Blombos Cave in South Africa that show abstract carvings.[41] The complexity of these designs suggests a significance to those peoples beyond that of everyday life, a significance passed on via language, as well as through observation and copying.

Soon after the time of the Blombos fossils, modern humans with their evolving stone technologies entered Australia 65 000 years ago, and on into the Americas from about 15 000 years ago. These migrations saw waves of extinction of large mammals and birds that continued with the human colonization of New Zealand and Polynesia in the last 1000 years. The sophisticated stone tool sets of these anatomically and behaviourally modern humans seemed to be an irresistible force in these newly entered landscapes: only in Africa, where humans had coevolved for countless millennia with these large animals, were the megafauna to survive in large numbers. Now, human information could be passed verbally through one generation to the next, via cave art from Europe to Indonesia, to papyrus and paper, and now to silicon. Such amassed knowledge would ultimately enable technologies that would build new kinds of skeletons, skeletons that would reshape the landscape and also life on Earth.

The vertebrates are not the only group with an internal skeleton. There are other groups to explore: ones that have an entirely different skeletal blueprint.

The 'Hedgehog Skin' Animals

Alongside the obscure vertebrates of the early Cambrian seas were another group of animals making endoskeletons. These are called the Ambulacraria, which includes two important animal groups: the hemichordates, which comprise the pterobranchs and acorn worms; and the much more familiar echinoderms, the group that includes sea urchins, starfish, and sea cucumbers. The last common ancestor of this diverse group of animals seems to have been some kind of bilaterally symmetrical filter-feeding animal.

Acorn worms do not build skeletons, while the elaborate 'colonial properties' of the pterobranch graptolite skeletons have already been noted. Echinoderms build skeletons in a myriad of designs, the name itself meaning 'hedgehog skin'. All echinoderms have skeletons made of calcite, though in the sea cucumbers—widely harvested as food under names such as beche-de-mer and trepan—most of the biomineralized skeleton has been lost. The bilateral ancestry of echinoderms is still evident in the larvae of modern echinoderms. But as they grow their body symmetry becomes radial, or more correctly pentaradial—fivefold.

At first sight, the skeleton of an echinoderm looks much more like an exoskeleton. But the hard skeleton of echinoderms actually originates from the 'middle-layer' (mesodermal) tissues of the animal, just as with the skeleton of a frog, rabbit, and human. Each of its skeletal plates and spines is made of a single crystal of the mineral calcite, while their external shapes can take various complex forms. The plates are perforated to reduce their overall weight, and they are enclosed within a layer of epidermis, so that the skeleton is truly beneath the skin. In some echinoderms, such as starfish, the plates are more loosely bound, allowing a degree of flexibility as the animal moves. In others, such as the sea urchins, the plates are more tightly configured as continuous internal armour. Inside the skeleton is a complex set of structures that include a gut and reproductive organs—the latter considered a delicacy in Japan. The skeleton of echinoderms provides the support for the many tubular 'feet' of the animal, which are maintained by a hydrostatic water vascular system, which the animal uses to roam around the sea floor.

Echinoderms are a successful group. Whilst not reaching the dizzying heights of a million insect species, there are a respectable 7000 living types and over 10 000 fossil species known. These animals are essentially sea dwellers, though some stray into seawater-influenced estuaries and lakes.

Modern echinoderms have fivefold symmetry as adults, but some of their most ancient relatives were animals with a bilateral symmetry that grubbed around on the Cambrian seabed. *Ctenoimbricata*[42] is one of these ancient animals that lived in the seas of France and Spain more than 500 million years ago. It had a disc-shaped skeleton of calcite plates, its upper surface covered with spines, giving it the overall appearance of a formidable boundary fence.

Why then if they began as bilateral animals, with a clear sense of left and right, have echinoderms evolved to be radial? As their larvae settle, they begin to lose this sense of left-ness and right-ness, the left side developing to become the oral surface with the mouth, whilst the right side becomes the opposite, aboral surface. An adult starfish, when it moves across the seabed, sometimes leads with one arm, sometimes with another, not quite sure which 'foot' it should put forward first.

As the planktonic larva of echinoderms settle, and attach at the seabed, they undergo their metamorphosis into adults. In echinoderms the larvae attach at the anterior (front) end, and as a result, the mouth faces down. Therefore, it must rotate through 90 degrees to face away from the mud of the sea floor. It seems to be this effect—the rotation of the mouth—that echinoderms have tried to accommodate by making their skeleton radial. Echinoderms, then, appear to be rather confused animals, not sure where there top or bottom is, or whether they should be walking forwards with one arm or another.

Some echinoderms have tried to resolve this confusion by trying to become bilateral again. There are the so-called 'regular' sea urchins, which conform to a radial pattern. And then there are the so-called 'irregular' sea urchins that have superimposed a bilateral symmetry on this radial pattern. 'Irregular' sea urchins probably don't think of themselves as such, particularly as they form evocative heart-shaped patterns, such as the famous fossil sea urchin *Micraster*. It is a pity Charles Darwin did not look closely at *Micraster*, which lay literally beneath his feet. It would have made him much less downbeat about what palaeontology can say about evolution.

The Shape-Shifters

Charles Darwin brilliantly divined the fundamental mechanism that drives the formation of coral atolls. For most scientists, such an achievement would be enough of a legacy for a lifetime's work. Darwin, though, was famously (and notoriously, still, in some circles) to go on to tackle the yet more fundamental question of how all living organisms took on the 'endless forms most beautiful' that they possess.

His 1859 bombshell on this subject, On the Origin of Species, is where he sets out this story. It is all put down somewhat more hurriedly than he would have liked. After many years of procrastination and piling up evidence, Darwin was pushed into making his definitive statement on evolution[43] by the news that Alfred Russell Wallace had arrived at the same idea, and was hot on his heels. In the Origin, a mass of evidence is set out to show how animals and plants can change their form through time. To a geologist reading the work today, though, what is striking is how little faith Darwin had in the fossil record. The title of the tenth chapter summarizes his views: On the Imperfection of the Fossil Record. What he wanted, at the time he was working on this problem, was to see fossils showing chains of intermediate links. But the strata to him did not seem to show this, and he complained of the 'paltry display' that we see even in 'our richest geological museums'.

One can understand his diffidence at the time. His colleagues were then writing furiously on the many new discoveries of fossils being made— some humdrum, some astonishing, some bizarre. But a glance at the geological journals of the time will show a kind of palaeontological lucky dip, with learned articles on all kinds of individual marvels: a new fossil insect, some finely petrified elephant bones, fossil footprints found next to salt-bearing strata, the teeth of extinct sharks, a new kind of plesiosaur, mummified cats. This clearly showed that the life of the deep past was full of strange and wonderful organisms that were no longer alive, but what it did not do then was to put them into systematic lineages that would show, specifically, how ancestral forms of life gave rise to new and different species.

Darwin need not have worried. The next century was to provide abundant examples of evolutionary lineages, as new generations of palaeontologists turned away from the scattergun approach of the pioneers, to more systematically study the fossilized skeletons that they were hammering out of the strata, layer by layer. Some classic examples were to come out of the very chalk strata that Darwin's home, Down House, was built upon. The meticulous researches, in the late 19th century, of that gifted amateur—a country doctor by trade—Arthur Rowe, on the fossilized sea urchins of the genus *Micraster* within the chalk, remains a textbook example of evolutionary change. These particular fossils, on ascending the strata, change almost imperceptibly from a flattened shape to an arched one, from narrow to broad, and from a mouth set back in the skeleton to one that slowly became more forwardly positioned. Rowe bemoaned the many different species names that had been given individual forms along what he saw as a story of gradual transition. For to him, along lineages, different forms seemed to grade into each other insensibly.

This was outright vindication of Darwin's theory—and outright settling of Darwin's doubts and worries about the fossil record. This painstaking, albeit amateur work on *Micraster* was noticed by no less a figure than the great science-fiction writer H.G. Wells, who together with his son, and the grandson of the great 19th-century naturalist T.H. Huxley, Julian Huxley, penned the influential 1931 book *The Science of Life*. In that book the fossils of *Micraster*, as collected by Rowe through those great thicknesses of the chalk rocks, are used as a classic study to demonstrate gradual evolutionary changes to the shape of a skeleton through time. In the case of *Micraster*, a sea urchin that was a burrower, it made sense over time to change the shape of its skeleton, to shift the mouth forwards and move the anus backwards so that it did not foul its burrow. This shows the inherent plasticity in the echinoderm skeleton, that though they are mostly radial as adults, they can adapt to a more bilateral morphology for moving directionally through a burrow. And *Micraster* was a veritable survivor, evolving many different forms, and surviving the calamity of the end-Cretaceous mass extinction too.

The echinoderms beautifully display patterns of skeletal evolution beyond demonstrations of gradual change in shape through time. The tooth-bearing apparatus of sea urchins is one striking evolutionary innovation. It is a structure that Aristotle marvelled at and described, and so it is now called the 'Aristotle's lantern' after him. The echinoderm apparatus is a broadly cylindrical array of about 50 individual skeletal parts, held together by about 60 muscles which manipulate the five parts that serve as teeth; these can be protruded out from the mouth to rasp at the sea floor. It seems an extraordinary complex means of making teeth—more like something out of Heath Robinson's fertile imagination than an outcome of biological evolution—but it has clearly been effective for sea urchins for many millions of years. For good measure, the teeth in some species are venomous.

Some other echinoderms have coevolved their skeletons to forge bizarre relationships with other skeleton-bearing animals. The elegant 'sea lilies'—more correctly known as crinoids—are mostly sedentary echinoderms, fixed by a stalk to the seabed. They have a long history, first appearing in the seas of the Ordovician Period, and living in modern ocean depths from the shallows to nearly 10 kilometres below the sea surface. Above the armoured stalk of the crinoid, the animal bears a 'calyx'—a skeletal crown of calcite plates, which houses the mouth, gut, and anus. From the calyx the crinoid unfurls its elegant, armoured tentacles, which sway in the current to extract small particles of food. Long ago, during the Ordovician Period, some small marine snails called playceratids took advantage of the crinoid's elegant crown, and can be seen regularly fossilized clinging to their hosts.[44] Whatever this relationship was, it must have been successful as it continued on until the Permian. When this association was first commented on in the mid-19th century, the snails were thought to be in the process of being consumed by the crinoid. But the relationship is, in a way, the converse. Careful examination showed that the snails are always positioned over the anus of the crinoid and seem to have chosen a diet as 'poo eaters'. Darwin would have thought it just another example of the fashioning of his 'endless forms most beautiful'.

4

PLANT SKELETONS

General Sherman weighs more than 2000 tons, is nearly 84 metres tall, and may be nearing 3000 years old (Figure 23). This is General Sherman the giant sequoia tree, or *Sequoiadendron giganteum*, currently living in California, and not General William Tecumseh Sherman, one of the notable commanders of the American Civil War. General Sherman the tree was named after General Sherman the general in 1879 by the naturalist, formerly lieutenant (in Sherman's army), James Wolverton, a few years after another sequoia tree had been named after General Ulysses Grant, who had been Sherman's commander in that war. It is a neat reversal of roles. General-Sherman-the-tree is the world's largest tree, and General-Grant-the-tree is (by a very small margin) the world's second largest. Both these general-trees are the largest by bulk, and not the tallest of trees. That distinction is owned by a redwood, *Sequoia sempervirens*, named Hyperion, also in California, over 115 metres tall, and presumably named after the Titan of Greek mythology of that name—and thus also a tribute to military force.

These trees are certainly giants, and an entire order of magnitude larger than the largest animal that ever lived, the blue whale, which reaches a mere 30 metres long and 180 tonnes in weight (Figure 7). It is a measure of one way in which the land plants have out-scaled the animals, despite being a late arrival on the planet.

Are we talking about *skeletons* here, though? After all, we are now in the realm of plants, and not of animals that have mineralized bones, teeth,

Figure 23. General Sherman.

and shells. Nevertheless, the matter of trees and their kin represents rigid, biologically built structures that are so commonly fossilized that they provide us with an energy source—coal—potent enough to power much of our lives for the present, and in doing so to imperil our planet's climate in the near future. We can consider the substance of trees as essentially skeletal.

No matter that much of a tree is dead tissue. Its construction allows such enormous statures to be attainable. Cut open a tree trunk and one commonly sees the annual rings of the wood as it grows outwards from the centre, the rings marking the difference between faster summer growth and slower winter growth. Looking closely with a hand lens or, better, a microscope, and the remarkable tubular pattern of the cylindrical water-carrying cells becomes apparent. These are made of a combination of cellulose fibres (which resist tensional forces) embedded within a mass of lignin (which resists compression). Those of the outer parts of the trunk, termed the sapwood, are functional, while the inner part of a large tree is of darker heartwood—something of a misnomer as this is dead tissue that the tree can live without, if the centre of an old tree has decayed away. Altogether, these tissues build up an impressive and durable skeleton.

The tree Hyperion must lift water over a hundred metres to its leaves high in the sky, via a system of cellular pipework and the force of nothing more than the evaporation of water from those leaves, in the process of transpiration. Then, it must translocate the sugars and other nutrients it has formed in those leaves, by photosynthesis, to the rest of the tree via another cellular pipework system. And it must do this year by year, while the structure of the tree is tested by gravity and by high winds. The skeletons of Hyperion and of the tree-generals Sherman and Grant have enabled this process within their own organisms for many centuries.

Some individual trees have carried out these processes, year by year, for longer still. Some of the twisted, hunched, and (barely) living skeletons of bristlecone pines, *Pinus longaeva*, have withstood the high altitudes, aridity, and poor soils of the White Mountains—also of California—for over 5000 years. The harshness and poverty of the environment has excluded most plant competitors, while a cocktail of resins in the wood has slowed water loss and discouraged the ravages of both decay and insects. Ironically, these conditions seem to have shortened the lives of the human researchers into these most ageless of organisms. The botanist Edmund Schulman, for instance, who more than anyone established the great age of the bristlecone pines, succumbed to a heart attack at the age of only 49. He drilled the core sample of the (currently) record-breaking tree, but did not live to see the results. The man who counted the tree-rings of this core, though, to establish the age of the tree as 5066 years old, Tom Harlan, bucked this unlucky trend: he lived well into his eighth decade. Ironically, the bristlecone pines themselves may not enjoy similarly extended longevity, as conditions in their specific habitat are now being threatened by climate change.

These are the most spectacular examples of size and age. But, even now, much of the world remains clothed in thick stands of trees. These structures set the boundary conditions for both the physical and the biological world within them—and deforestation alters those boundary conditions profoundly. The forested world is often held to be a primeval one, and, true, it stretches back a few hundreds of millions of years. Nevertheless, by comparison with the innovations of the Cambrian explosion, it is a newcomer. The sylvan ancestors of Generals Sherman and Grant show that forest skeletons began on much more modest scales.

Early Days

What took the plant world so long to take over the land? There is no easy answer to this simple question as yet—a question made more difficult by the fleeting and indirect glimpses offered to palaeontologists of the early and emphatically non-skeletonized pioneer forms. Simple plants have thrived in the Earth's oceans for more than 3 billion years, but the first indication that plants—as a green algal scum in some moist and sheltered places—probably established a toehold, and perhaps more, on land is seen in rocks over 800 million years old. This cryptic invasion has been inferred from deductions made that biological weathering of land surfaces began about then.[45] Sometime in the Cambrian and Ordovician, this green algal scum probably became more organized, to form filamentous algae perhaps resembling the stoneworts of today, which typically occupy shallow freshwater pools.

Then, in rocks some 470 million years old, laid down in the early part of the Ordovician Period, there are the first hints of a resistant plant skeleton, in the form of minute spores tough enough to survive burial in sedimentary rocks, and subsequent extraction by palaeontologists using strong acids. These spores resemble those of modern liverworts, which comprise a 'thallus'—sheets of plant cells, leaf like in outline, but lacking veins or stems, and being attached to the ground by root hairs rather than by organized roots.

In rocks of the Silurian Period, about 430 million years ago, we find the first evidence of a plant that had begun to raise itself off the ground. This was *Cooksonia*, named after the talented Australian botanist Isabel Clifton Cookson (1893–1973), who gained her scientific reputation for work on plants both modern and ancient. 'Cookie', as her colleagues knew her, collected specimens of some of the world's oldest plants from the rugged terrain of the Yarra River. While visiting Manchester in England, she struck up a long-lasting collaboration with William Henry Lang, a professor at the university there. Lang had been studying medicine, but somehow emerged as a palaeobotanist, and his discovery, *Cooksonia*, became a tribute to his antipodean colleague. Cookie later became a shrewd investor in the stock market, and used her winnings to fund her research in retirement, when she wrote 30 of her 85 papers. She organized this work, though, so that she could faithfully listen each week to radio broadcasts of *Blue Hills*, a long-running serial on Australian farming folk, and the iconic, and ambitious, children's programme *The Argonauts*—which perhaps reminded her of her own childhood ambitions, being pursued with vigour to the end.

Cooksonia was a plant that could be said to be the polar opposite of a liverwort, being essentially just a branched, photosynthetic stem, a millimetre or so across and a few centimetres tall, anchored into the ground. It had no leaves, nor genuine roots, but it did have spore-forming bodies at the ends of some of the stems. And some forms of it possessed another innovation, whereby strings of cells within the stem died to become hollow tubes that could carry water around the plant. This was the beginnings of the vascular system possessed by advanced plants, and that were to become colossally extended in Generals Sherman and Grant, and yet more so in Hyperion.

Plants robust enough to become properly woody, and so to develop a respectable skeleton, were to emerge in the subsequent Devonian Period. What is the oldest tree? For a long time, this honour was generally held by *Archaeopteris*, which resembled something like a mixture of conifer and fern, up to 10 metres high, and which was widespread in the later part of that period, about 360 million years ago. But a few fossil tree-like stumps in a famous 'fossilized grove' at Gilboa, in New York State, had been an enigma since they were first discovered in 1870. These dated from earlier times—the middle part of the Devonian Period—but little more was known about them. They were assigned to the genus *Eospermatopteris* and left in the 'pending' category of tantalizingly incomplete specimens so familiar to palaeontologists.

Then, in 2005, a spectacular, complete specimen was unearthed at Gilboa.[46] It was attached to fossil fronds that had previously been described under the name *Wattieza*, and together the two made up a tree with fern-like fronds, rather than leaves, that could reach 8 metres in height, with the name *Wattieza* being retained because it had been named first, and that of *Eospermatopteris* being discarded. The picture of the world's oldest tree (so far) was complete. It was only distantly related to a modern tree such as an oak or sycamore, but rather belonged to a group called the cladoxylopsids, which were closely related to today's ferns and horsetails, and which reproduced by spores, not seeds. A tree skeleton, therefore, is an engineering solution that may be arrived at by more than one route.

By the mid-part of the Carboniferous Period which followed, around 300 million years ago, there was an explosion of forests with yet more outlandish tree-sized ferns, horsetails, and other—to our eyes—bizarre-looking 'trees' that could reach 50 metres in height. As with *Wattieza* and *Eospermatopteris*, there have been conflations of identity, as different parts of fossilized plant skeletons have been found and named. Impressive, root-like structures called *Stigmaria*, for instance, are not uncommon as fossils in Carboniferous strata—and these were later found to be the underground parts of fossilized trunks that had been given the names *Sigillaria* and *Lepidodendron*. The latter has a distinctive ornamentation of diamond-shaped leaf scars, which some people thought had a very reptilian appearance—and so in the 19th century, examples of these used to be put on display as giant fossil snakes!

The coal seams that these bizarre trees contributed to have long been a key factor in geology, not least for their importance in providing the energy to spark, and then to drive, the Industrial Revolution. In France, when the Comte de Buffon was devising arguably the first scientifically based Earth history in the late 18th century, *The Epochs of Nature*,[47] coal seams were one of the key examples analysed by this industrious and perceptive savant of France's *ancien régime*. He was perfectly aware that they were the compressed and buried remains of long-dead forests, and in a *tour de force* of deductive reasoning he compared these not to modern French forests, but to the tropical coastal swamps (that he vividly described from travellers' accounts) of Guyana—an analogy that still holds true today. A little later, coal seams—and the important question of where one may find them underground (and become rich) or not find them (and become bankrupt)—provided one of the main strands of William Smith's development of the sciences of both litho-stratigraphy and biostratigraphy in the early 19th century, and of his almost unbelievable feat in making the first geological map of Britain—single-handed.[48]

This extraordinary forest expansion of the Carboniferous still casts a long shadow both in the science of geology and in the global economy, and in the environment, today. The abundant coal seams it gave rise to are in part a freakish accident of geology, when the evolution of trees with a proper woody skeleton coincided with a spread of low-lying swampy conditions in the tropical zone virtually across the entire world, from North America to Britain and Europe, to Asia as far as China. These conditions were perfect to not only grow such a mass of trees, but to preserve them in the boggy ground—and then to bury them deeply to be transformed by heat and pressure into coal.

The emergence of the late Devonian and Carboniferous forests had a number of consequences. One was the absorption of large amounts of carbon into woody tissues and burying it underground once those wooden skeletons fell to the swamp floor. This was an effective form of carbon sequestration that diminished the Earth's greenhouse effect, and lowered global temperatures so that, far away from the tropical woody swamps, a massive ice sheet grew around the South Pole, plunging the Earth into a glaciation that lasted, on and off, some 50 million years. Ironically, this ice age, and the global rises and falls in sea level as the ice repeatedly advanced and retreated, was another factor in forming and preserving the coal seams. Each sea level fall exposed large areas of coastal plain to become forested, while the subsequent sea level rise drowned the forest, which was then covered in thick layers of marine mud to begin the process of fossilization.

As the tree skeletons grew and were entombed, and large amounts of carbon were stored underground, the effect was not only on climate. There was a progressive accumulation of that by-product of photosynthesis, oxygen, in the atmosphere, and its levels seem to have risen significantly higher than those today. One effect of this was to make forests more flammable, even under wet swampy conditions, and fossil charcoal is one of the components of the coal material. A high oxygen content, too, may have been a factor behind the gigantism of some of the Carboniferous arthropods—that extraordinary 2.5-metre-long millipede *Arthropleura*, and the dragonflies such as *Meganeura* with 65-centimetre wingspans. One kind of skeleton—in this case by its mass burial—can clearly help another.

An arms race took place among plants that saw them grow ever higher to reach the sunlight (and hence overshadow their smaller neighbours), and develop ever more effective ways of preventing water loss from their tissues—for instance by developing tough waxy outer layers that would allow them to colonize ever higher and drier regions.

With the plants, there came the animals. For the plants, big and small, built up to be a gigantic store of carbon, and nitrogen, and phosphorus, and other life-giving elements—that was food for animals, and these, the herbivores, in turn became food for the carnivores. The animals' arms race was interlinked with the arms race among plants, almost from the beginning of life on land. The many battles in this fight for life took place in the complex three-dimensional spaces afforded by the various levels of forest canopy—spaces that were insulated from extremes of heat and cold and drought by the shelter of the leaves and branches, by the stores of moisture and humic matter in the soils beneath, and by the rains produced as a result of billions of trees transpiring. The forest skeleton is, in truth, the main cradle of life on land.

There were yet wider consequences of plant skeletonization, though. This was the armouring of the land itself, and a change in the way that rivers flowed.

The River Revolution

For most of the Earth's history, much of the land was inhospitable—a lifeless wilderness mostly made of rock, scree, and sand dunes. As sentient animals, we react to the 'lifeless' part of the description, but think less about the shape of the land, perhaps assuming that such underlying terrain is much the same whether or not it is covered by a thin living skin. However, that seemingly delicate and transient organic skin is skeletonized, and so has a surprising power to change the structure of, and not just clothe, the physical landscape.

While each river is unique in its particular outline, rivers as a whole can be grouped into a few broad categories, which can effectively summarize the way that currents of water, under the influence of gravity, can flow across a land surface. The rivers that many of us, living in mild and temperate landscapes, are most familiar with are those where the flowing water is more or less confined within a single deep channel that is shaped into broad loops across the river valley. Watch such a river for long enough—a few years is usually enough to appreciate the pattern—and one can see that these loops are not fixed but move gradually across the river plain, with erosion of one bank being balanced by the build-up of river sediment on the opposite bank. Over time, the river channel can move hundreds of metres across the floodplain—and then back again. This kind of 'normal' (to many) river is a meandering river, sometimes described as a 'single-thread' river, because most of the water flows along a single channel.

The gold prospectors heading avidly for the Yukon and Klondike rivers in the 1890s picked their way across—and dug for gold into—rivers far removed from the meandering rivers of the temperate climes that they had come from. These were shallow rivers, with not one but dozens of shifting channels, where water might race along, or might just as easily quickly sink into the gravel, to leave the river dry. The settlers moving out across the Nebraska plains saw similar landscapes across the Platte River, disgustedly calling them 'a mile wide and a foot deep'. These are braided rivers, most typical of the bleak terrains of recently glaciated areas, or at high latitudes where glaciers still hold sway.

These different types of rivers can now be observed and analysed, and the flow of water and sediment in them can be measured day by day, and hour by hour. But to get an idea of how rivers operated in the Earth's deep geological past, it is no longer possible to examine the rivers themselves—they have long gone. Instead, one has to analyse their traces, in the form of the rock strata that they have left behind. Meandering rivers form specific patterns of strata, left as the river channel has played back and forth across the flood plain. A stratal sequence might start with a coarse gravel, representing a channel bottom winnowed by fast-flowing water. Above this there might accumulate a sand deposit laid down by the river bank building out over the channel, and then as the river channel migrated further away, there would be strata of mud, that built up on the flood plain as the river periodically broke its banks and swept across the floodplain. These muddy strata may have the remains of roots in them, from the plants that colonized the floodplain. When the river next migrated across this spot, hundreds or thousands of years later, this 'fining-up' sequence would repeat itself, and this particular pattern—often repeated many times in thick successions of strata—is now firmly linked in geologists' minds with the action of an ancient meandering river.

Braided rivers, with their more irregular flow and multiple channels, leave different kinds of strata. The strata here are dominated by irregular beds of gravel and sand, representing the channel bottoms and the river bars that separate them. There is not much here in the way of the thick muddy deposits that are preserved on the floodplains of meandering rivers, for the whole river here is a kind of irregular, ever-changing flood plain, and mud is quickly swept farther downstream.

In practice, of course, the distinction is not always so clear cut. There are other kinds of strata, such as those that form near coastlines, on estuaries, beaches, and tidal flats, that in some respects can resemble the strata formed by rivers. And meandering rivers and braided rivers are not immutable categories, but can change from one style of behaviour to the other. Many of the classic meandering rivers of temperate latitudes today, for instance, behaved as braided rivers in the cold climates of the Ice Ages. Nevertheless, examination of river-laid strata back into the deep past of the Earth is now showing a clear pattern: in the ancient times of the Precambrian and the early part of the Palaeozoic Era, from over 3 billion years ago to about 430 million years ago, most of the Earth's rivers behaved like braided rivers.

Then, in the Silurian, Devonian, and Carboniferous, river patterns changed. River channels began to commonly take single-thread patterns, while the adjacent floodplains stabilized, and began to accumulate large amounts of mud. The critical factor has been identified as the grip that the growing forests took on the landscape.[49] One might think of it as a kind of armouring of the land, arising from the collective skeletons of the trees. Their root systems ramified through the loose sediment on the ground surface, holding it in place and allowing clay and silt to accumulate, and dead, decaying plant matter also, so that thick humus-rich soils could develop. These transformed surface sediment layers were now heavy and cohesive, and further held together by the ramifying root systems. Hence, it was no longer so easy for rivers to split into a multiplicity of channels that could change their pattern through the loose surface sediment with every flood. The river waters were increasingly forced, especially in the forested lowlands, into single channels that could only gradually evolve their outline as the meander bands migrated, decade by decade and century by century, across the now-armoured floodplain.

This was a revolution in river behaviour—the second major one recognized in Earth history. The first had taken place 2 billion years previously, as the atmosphere began to fill with oxygen and the river minerals changed from reduced to oxidized forms. A third great revolution in river behaviour is now in progress, driven by humanity's direct and indirect effects on the landscape. *That* is really another story, to come to later.[50]

The Moving Plant Skeletons

We are used to a world of mobile animals that can run, and leap, and fly swiftly through the air—against a tableau of immobile grasses and trees, that do little more than bend to the wind. Why such a contrast in behaviour? If trees, with their massive and tough skeletons, were as mobile as animals, they could be unassailable competitors to animals, rather than providers of both nourishment and of shelter, and indeed providers of a complex three-dimensional environment where animals can thrive and prosper, much as corals underpin and shelter the complex and diverse ecology of a reef system.

There are ancient legends of walking trees, such as the Japanese tree spirit Ki-no-o-bakė, that could both walk and assume different guises. There are some modern legends too. The 'Walking Palm', *Socratea exorrhiza*, has been claimed to be able to 'walk' as much as 20 metres a year. It is certainly a strange tree. Its name *exorrhiza* means 'outside root' and it possesses what are called 'stilt roots', which can extend a metre or more above ground, before the trunk begins. It has been said that this palm can, by growing new roots on one side, and suppressing them on the other, shift its position progressively so as to take up, say, a better position to receive sunlight. It is a lovely story but—alas—is probably untrue.

In reality, there are some highly mobile plants—of a kind—but they are very small. The tiny, single-celled planktonic dinoflagellates, propelled through the water by means of their whip-like flagellum, are mostly photosynthetic, but they can also be predators of even smaller prey. A little larger is the extraordinary *Volvox*, where up to 50 000 individual single-celled algae—each armed with a couple of flagellae—form spherical colonies that can collectively swim towards light.

Plant skeletons, though—or parts of them—can adapt to be carried by the wind. The iconic example is tumbleweed, those rolling masses so often used to symbolize the bleak and pitiless landscapes against which westerns are filmed. And indeed there is the classic 1953 western *Tumbleweed*, which is not to be confused with 1999 film *Tumbleweeds*, where the symbolism is used to depict failed human relationships, or the 2012 film *Tumbleweed!*, which is mysteriously described as the true and historically accurate story of the tumbleweed that would not tumble.

Tumbleweed might be used to symbolize death and resurrection, too. The weed that tumbles is the plant that detaches itself from its roots, then to be blown along by the wind. Its purpose is simply to spread and reproduce the species, and it does this best in savannah or semi-arid environments—like the North American mid-west—where there are not too many living trees to stall its progress. The only living parts of tumbleweed are the seeds or spores that are being carried along the ground. The carrier is a husk, a dead skeleton that only has the purpose of catching the wind. In this functional death, the spores or seeds spill out during tumbling, or are released once the skeleton has come to a halt.

It is an effective lifestyle—or perhaps deathstyle—for a plant, and so there is not just one kind of tumbleweed but many, with at least ten major groups of plants developing this habit. It is a habit, too, that travels beyond even the reach of the wind. One might consider the genus *Kali*, in the family of the amaranth plants. Kali in Hindu religion is a goddess, known as a destroyer—though of evil rather than of good—and is often portrayed dancing on the prostrate form of her consort, the god Shiva. *Kali* the plant might be held by a botanist to have comparable powers. It is a Eurasian plant by origin, the 'Russian thistle', but when it entered North America in the 1870s, via a shipment of what should have been flax seeds, it found the mid-west to its liking and took hold. It soon became an omnipresent naturalized species that, when Hollywood's camera lens started rolling for the cowboy adventures, became the eternal symbol of rootlessness, blowing through that wide landscape.

Tumbleweed is not just a North American phenomenon, but (often as the invasive *Kali*) is present in Central and South America, and southern Africa, and Australia. Its rolling can have spectacular, and sometimes disconcerting, effects. In 2016, the Australian town of Wangaratta was invaded by a mass of the tumbleweed *Panicum effusum*, which in antipodean vernacular became 'hairy panic'. The hairy panic piled up so thickly in places—to several metres high—that residents could not get into their houses, or find their buried cars. The local council, though, had a secret weapon at its disposal—it would, it said, arm its street sweepers with vacuum cleaners. History seems not to have recorded the consequences.

Plants do not need to have their dead skeletons blow wholesale across the landscape to help their reproduction. They can simply have those microscopic yet heavily armoured capsules of reproduction, their pollen, being blown large distances cross-country instead. Pollen is technically the male gametophyte of plants, which might be thought of as plant sperm, designed to be produced in massive numbers for some to land, by chance, upon the female gametophyte—the pistil of a flower—of another plant of the same species. It is a game of chance, that the plants try to stack in their favour by going into productive overdrive, releasing—as hay fever sufferers know—these grains in countless millions (though many plants have devised a means to load the dice, by tempting insects to spread their pollen for them from plant to plant).

Pollen grains have an impressively tough outer coating, tough enough to withstand the action of strong acids in the laboratory, while the mineral grains around them dissolve. It is so tough, indeed, that biochemists have not fully succeeded in breaking it down to know exactly what it comprises. They have, though, given it a name, sporopollenin, and detected within it a suite of complex and interlinked biological polymers. The outer surface of the sporopollenin is sculpted, too, into a variety of intricate patterns, each specific for each plant species (Figure 24). This has been one key to working out forest histories, as we shall see.

The distances that pollen can travel are extraordinary. Pine pollen has evolved wings, which make each grain look in outline a little like a silhouette of a Mickey Mouse head (Figure 25). It can look amusing

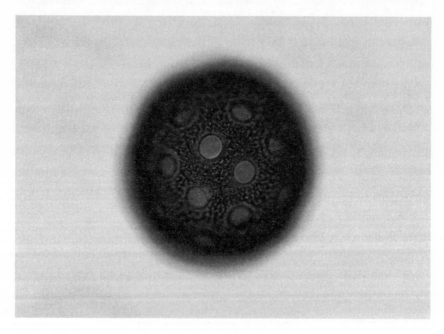

Figure 24. 'Golf ball' pollen, *Agrostemma githago*, diameter 56 microns.

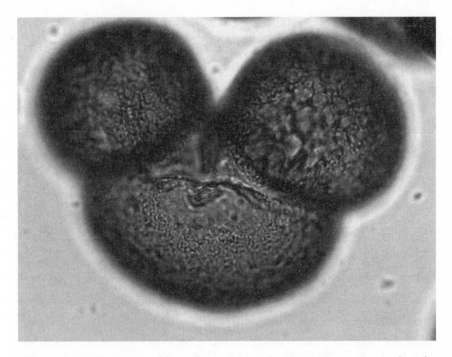

Figure 25. Mickey Mouse-like pollen, *Pinus nigra*, 75 microns at greatest length.

(to humans) but is highly effective as a means of transport. Pine pollen grains have been known to be blown at least 3000 kilometres—and can still germinate after a flight of at least a few tens of kilometres.[51] Pine is therefore a little notorious in the annals of pollen palynology, which is the science of reconstruction of the nature of prehistoric forests from the pollen that drift from them, eventually to land within peat bogs, or in lake beds, or that drift on to the sea floor, for these tough little time capsules to be fossilized within the accumulating strata.

Over geological time, the nature of forests has undergone repeated changes. This was particularly so during the Ice Ages, when the mid-latitude regions of the Earth oscillated between balmy warmth—when oak trees flourished—and bitter cold, when the oaks died and pine or birch, or the grasslands of arctic steppes, took their place. There are rarely enough fragments of the trees or grasses themselves preserved to yield any sensible account of these complex forest histories. But the almost indestructible pollen grains leave a memory of how the forests changed. Extracted using acid from their enclosing strata, they may be interrogated to work out how vegetation, and therefore climate, changed through time. The pollen record is skewed, though, for some plants produce more pollen than others, and some pollen—like that of the pine—can be blown farther than others. An incautious palynologist might reconstruct a whole forest from the pollen produced from a few lonesome pines.

The Grit in Grass

When battle was about to start at Troy, Zeus sent blood rain from the heavens as a warning of slaughter to come. When Sarpedon, his own mortal son, was felled in that slaughter, then Zeus once more showered the Earth with drops of blood, this time through grief.[52] These passages of Homer's *Iliad* are the first written mention of this eerie rain, which was later quoted also variously by Hesiod, Plutarch, Livy, and Pliny, usually as a portent of evil to come. Later, in medieval times, blood rain was seen as a harbinger of the Viking invasion, of the exploits of Richard the Lionheart, and of the Black Death. Up to the 17th century, people thought that the red-stained rain that fell from time to time was literally a fall of blood. Then, slowly, people began to seek natural causes for this phenomenon.

So, when in Lyon in France, on the 17th of October, 1846, there was a violent gale accompanied by showers of blood-red rain, the scientist Christian Gottfried Ehrenberg took an interest. Samples of the dust that remained when the rain dried out were sent to him, together with a melodramatic eyewitness account of the frighteningly dark skies, stifling winds, bolts of lightning, and the maddened birds in the fields, so bewildered that they could easily be caught. Ehrenberg's keen eye, and his microscope, solved the mystery. The dried 'blood' was, he said, just desert dust from Africa, blown in with the trade winds. But, in his celebrated account of the Lyon event, *Passat-Staub und Blut-Regen* ('trade wind-dust and blood rain'), he saw other things too. There were skeletons in the dust.

Ehrenberg saw in that material the characteristic pollen of pine—but there was much more besides. He recognized 73 species in total, and listed them all systematically. There were the silica skeletons of diatoms, and the calcareous remains of foraminifera, and the remains of *Gallionella*, an iron-oxidizing bacterium. These, he supposed, had been whipped up by the fierce winds from some dried-up ponds, or (for some of them were probably fossils) from some rocky landscape. Twenty-two of the species, though, were tiny, obscure skeletons—and he made something of a speciality of them. They were phytoliths, which roughly translates as 'plant stones'.

Phytoliths are microscopic mineral grains, normally a few tens of microns across, found within the tissues of many plant species. They are usually made of silica, though in cacti they are made of calcium oxalate. They are truly skeletal, in that they remain when the rest of the plant has decayed away, and often form the only record that it ever existed. Different kinds of plants make different kinds of phytoliths—many grasses have them, for instance, and provide a range of distinct phytoliths, which help identify what kind of grass contained them, while those from trees are typically less diagnostic as to the kind of tree.

Charles Darwin had noted what turned out to be phytoliths too, in circumstances similar to those of Ehrenberg's study of the Lyon blood rain, when examining dust samples blown on to the *Beagle* while it was near the Cape Verde islands, which he then sent to Ehrenberg. Ehrenberg, in turn, quoted Darwin, approvingly, in the first sentence of *Passat-Staub und Blut-Regen*—published a decade before *On the Origin of Species*—and corresponded at length with him. Later in life, though, Ehrenberg disagreed with Darwin about evolution. Microfossils, said Ehrenberg, did not evolve. Those were early days in the science. Microfossils are now known to provide some of the most elegant and convincing examples of evolution in the fossil record.

Why do plants form phytoliths in the first place? This is still not fully known. Phytoliths can help to support the structure of cell walls where they occur. In some plants they seem to afford some protection against infection, and they also make grasses harder and more abrasive, and therefore less palatable to those animals grazing on them.

Regardless of their function, phytoliths are now used by archaeologists and environmental historians to help reconstruct conditions of the past. Phytoliths have been recovered from between the teeth of human skeletons, and from archaeological hearths, to help establish what kind of diet ancient humans ate. They can be present in pottery, and such evidence has been used to help establish the complicated paths, for instance, by which maize became domesticated in the Americas, and how rice developed as a crop plant in Asia. It does not only occur in geologically recent material, having been reported from dinosaur dung, as a clue to what those animals ate.

Phytoliths are tiny. It is time to switch scale, to consider some of Earth's most enormous skeletons.

5

MEGA-SKELETONS

Humans have, uniquely among Earth's species, gazed across at their own planet from the vantage point of another solar body, the Moon. It is a sight that has, among many human individuals, stirred an appreciation of the Earth's complexity, and delicacy, and fragility—and of its hospitability to an abundance of life, by comparison with the barren surfaces of the planets and moons elsewhere in the Solar System. Images of the Earth from space may be used to track passing weather systems, changes in ocean currents, the growth and decay of ice, outbursts of ash and lava from volcanoes, and much more. They can also, even to the casual eye, reveal parts of our planet that are, in essence, masses of skeletons, left by countless organisms over almost unimaginable spans of time.

Some are easy to see. Pale blue or white-flecked patches can be seen against the otherwise dark blue canvas of the Earth's oceans. One such extends along much of the east coast of Australia (Figure 26). Others lie to the east of the Central American Isthmus, notably a large patch around the Bahamas islands. Other smaller examples are dotted in the Pacific Ocean.

The pale blue colour comes from the shallowness of the sea, only a few metres deep at most over areas that may cover thousands of square kilometres. The edges of these pale blue patches are frequently sharp, as their just-submerged edges plunge sharply down into deep water. In places, pale blue patches that are isolated in the middle of the ocean may have tiny rims of white or brown or green, as the shallow seas break the

Figure 26. Part of the southern portion of the Great Barrier Reef adjacent to the central Queensland coast. The reef is up to circa 200 kilometres from the coast.

surface to form strings of islands, some more or less barren, others covered with luxuriant vegetation.

These are the surfaces of monstrous mounds of limestone that often are little more than close-packed skeletons of generation upon generation of organism. They are the coral reefs and the larger structures, the carbonate platforms, which can extend from them, and which can include a multiplicity of skeleton structures of their own. Coral reefs are a byword for the profusion of diverse and colourful undersea life, and represent for biodiversity in the oceans what the tropical rainforests do on land. They are now, too, critically endangered biodiversity hotspots, which may well not survive as functional entities to the end of this century, being pushed to the brink by the multiple pressures that humanity is increasingly exerting on them. Living or dead, skeletons form the foundation of these gigantic planetary-scale biological constructions. And if we do drive them to extinction, there may be some scant consolation in knowing that part of their skeletal foundations will remain as a memory of their living magnificence, probably to the end of our planet's history.

The Coral Skeleton

Corals have long had a reputation for being both beautiful and mysterious. When, in Jules Vernes' *20 000 Leagues under the Sea*, Captain Nemo invites Professor Aronnax for an underwater stroll through a Pacific reef (complete with oxygen-providing Ruhmkorff apparatuses), that thoughtful savant recalls that coral—'medicine to the ancients, jewellery to the moderns'—was 'in the zoophyte branch'. Aristotle had coined the term 'zoophyte' two millennia back, for a class of organism he thought to be half-animal and half-plant. Aronnax, as he gazed entranced at the biological riches around him, further observed that only in 1694 did 'Peysonnel of Marseilles' establish that these bright, flower-like organisms were animals, not plants.

The good professor was correct about the man, though his memory was amiss as to the date. Jean-André Peysonnel in 1694 was in no position to make a major scientific discovery, as that was the year in which he was born. In 1720, he was a physician in Marseille, when the plague struck. Many people died, but Peysonnel's efforts on behalf of his stricken fellow citizens were so widely admired that he received a king's pension, and could retire to devote himself to his passion, marine science. He travelled to the Barbary Coast of North Africa, with royal instructions to record its natural history. He was aware—and suspicious—of the latest, highly influential treatise on corals by the Italian scientist Marsilius, who interpreted them as flowers—albeit flowers with petals that shrank back when touched. He resolved to test these ideas.

Peysonnel experimented on corals, dissecting them, observing their responses to different chemicals, and noting how dead specimens putrefied ('with a very disagreeable odour, approaching that of burnt horn'). The results convinced him that corals were 'insects'—that being the general name then for invertebrate animals. He prepared a paper detailing his evidence, and sent it to be read at the French Academy. In 1726, it arrived on the desk of its most influential member, René-Antoine Ferchault de Réaumur, regarded as the greatest living authority on 'insects'. Réaumur was wholly unconvinced by this new work, considering that Marsilius had settled the question unequivocally. Nevertheless, he agreed that Peysonnel's paper could be read at the Academy—but out of kindness, he suggested it should be read *anonymously*, for fear of public humiliation for the young physician. Réaumur's intuitions were well founded: there was indeed loud and immediate derision at the ludicrous idea that the magnificent coral groves could be produced by 'insects'.

Nevertheless, Peysonnel persisted in his views, and with a new position ('Royal Botanist' on the island of Guadeloupe in the Lesser Antilles) was well placed to extend his studies. New observations, together with the discovery that the related 'freshwater polyp' *Hydra* is an animal, convinced even Réaumur that Peysonnel's ideas were correct, and he was gracious enough to say so. Peysonnel, though, perhaps still feeling his fingers a little scorched from the Academy's initial reception, was to publish the rest of his work through England's Royal Society.

This early uncertainty about the nature of corals is understandable. Their morphology is relatively simple, without the obvious 'animal-like' tissues or indeed behaviour of, say, a worm or a crab. Most of their substance is a jelly-like material, the mesoglea, sandwiched between two thin tissue ('epithelium') layers. They have no head, no limbs, a loose meshwork of nerves, and only one mouth (which doubles as anus) into a combined gut and respiratory organ. But they do have tentacles (Marsilius's 'petals') around that mouth, which bear powerfully effective stinging cells. If you are of a certain age and have done a biology class, you may remember them as 'coelenterates', a phylum in which they were combined with ctenophores, the 'comb jellies'. However, ctenophores are now thought to be sufficiently distinct to be placed in their own phylum, which leaves the corals, together with the sea anemones, sea pens, jellyfish, box jellies, and hydrozoans (which include *Hydra*) in a phylum of what are now called the cnidarians.

The cnidarians have two kinds of body—the medusae, which are free-swimming, and polyps, which attach to the sea floor. Some species are just one or the other, while many alternate them in a life cycle in which the polyp reproduces asexually while the medusa is the sexual stage. In corals, the polyp is the dominant body plan, and in the main reef-building species, asexual reproduction allows the production of many individual, genetically identical coral animals within large colonies that may be up to a few metres across in size. For a sea floor organism to have the ability to surround itself with exact replicas of itself is a highly effective means of gaining and controlling space on the sea floor. When that possession is reinforced by the building of a solid rock platform to live on, then that territorial possession is rendered all the more *permanent*. Territory can be further expanded, too, during repro-duction, when eggs and sperm are released, combining to form a mobile larval stage that settles on the sea floor to form the first polyp of a new colony.

The coral rock platform is made up of the cup-like coral skeleton, which today is made of calcium carbonate in its mineral form aragonite, in which each coral animal lives. The individual cups, or corallites, interlinked in the vast building complexes of the coral colonies, are the key to the reef. They grow upwards, adding floor upon floor like the building up of a skyscraper. The coral animal itself is hoisted up in the process, so the skyscraper occupant moves up floor by floor, while the lower stories are abandoned. This construction adds the third dimen-sion to the territorial control that they had gained by the repeated asexual division. That third dimension, as we will see, can develop over time to be of huge extent. It is what ultimately turns biology into geology.

Geology students, much as they have to tell brachiopods apart from bivalves, also have to be able to recognize the major categories of corals, which are similarly common as fossils. The trick here includes recognizing which corals are solitary—essentially forming a single horn-shaped tube with an outer wall—and which are colonial, where the outer walls are conjoined in various ways, or can even be reduced or absent so that the inner structures can act as skeleton. These inner structures are the various successive 'floors' for the coral animal, with some floors being simple unitary structures, and some complex patchworks, while there is often a central column, often with radiating vertical plates, that can look from above like the spokes of a wheel. In whatever design, the skeleton shows radial symmetry to match that of the coral soft parts.

The corals grow upwards to reach the sunlight that is for them literally life-giving. For the reef-forming corals do not so much feed by using their tentacles and stinging cells to snare and kill passing small animals (though they do that too). Rather, they obtain much of their nutrient from microscopic passengers. These are *zooxanthellae*, specific forms of dinoflagellates, photosynthetic single-celled protists that are symbiotic with coral animals, gaining shelter and respired carbon dioxide from them, and in turn providing nutrients. Crowded within the coral tissues, they are what give the corals their brilliant colours, and what power their skeleton-building.

On a reef, the colonial coral skeletons take various forms. Some are rounded and ball like, while others, like the staghorn and elkhorn corals, take fantastical branching shapes. These shapes partly are the result of different species, which tend to produce different modular patterns in building their skeletons. Partly, though, it reflects the corals' sensitivity to their environmental surroundings, with some species growing shorter, stubbier colonies in areas where waves frequently crash in, and more delicate filigree patterns in calmer waters.

Billions of these combined skeletons make up the great, planetary-scale reef structures of Earth. The Great Barrier Reef—a complex structure of hundreds of islands and thousands of individual reefs—stretches for 2300 kilometres along Australia's eastern coast. Coral reefs spread across the tropical and subtropical belts of the Pacific, Indian, and Atlantic oceans. Their total area, globally, though, is small—less than a third of a million square kilometres, perhaps one-tenth of 1% of the ocean surface. Nevertheless, the biodiversity that they harbour is legendary, with perhaps 25% of the oceans' species. The intricate, stony, three-dimensional coral structures act as framework and shelter for a dazzling array of organisms, and these organisms in turn help shape the reef structure. Some of these interactions were glimpsed by an adventurous young scientist, as he travelled the world to describe, and think through, what he saw and what he collected. It was Charles Darwin's first major project.

The Wider Reef

When, in 1836, Charles Darwin visited the 'Keeling or Cocos atoll' in the Indian Ocean as part of his 5-year voyage on the *Beagle*, he was an energetic young man of just 27 years—a far cry from the elderly bearded sage that is now his usual image—and so was able to explore this remarkable island to the full. The double name of the island derives from its original discoverer, William Keeling, a captain of the East India Company (now best known not for this newly recognized piece of land, but for staging the first recorded amateur production of any Shakespeare play, which was of *Hamlet*, on his ship off the coast of Sierra Leone in 1607) and from the coconut trees which grew abundantly there. These Cocos—or Keeling—Islands are not the same as the Cocos Islands of the Pacific Ocean, where fabulous hoards of pirate gold were supposed to lie buried. Darwin, in any case, was focused on the scientific treasures on view, as he puzzled how this thin, circular coral reef island—just a few hundred metres in width—could enclose an equally circular shallow-water lagoon nearly 16 kilometres across (Figure 27).

In the days before aqualungs and submersibles, Darwin had to do his best to see from land the processes that he knew were taking place under water. From the island itself, just a few metres above sea level, he tried, at low water, to get to the rounded masses of the living coral in the sea

Figure 27. The Keeling Islands from space. For scale, the dimensions of the lagoon situated between the main island complex in the south of the picture are about 10 km by 16 km.

beyond. By use of a 'leaping-pole' he could reach coral masses where the tops of the coral colonies, just poking out of the water, were dead, while the sides, permanently in water even at low tide, were alive. Peering even further from this vantage point, he could see 'during the recoil of the breakers' that the fully submerged masses were wholly covered in living coral. These corals grew up to the sea surface, and could grow no farther upwards, so had to stop growing, or expand laterally.

These large rounded corals could resist considerable wave action, but even they could not resist the full force of the Indian Ocean breakers under storm conditions. Darwin saw that there were 'other organic productions, fitted to bear a somewhat longer exposure to the air and the sun'—and also better fitted to withstand storm force waves too. These he described as 'one of the lowest classes of the vegetable kingdom' that he referred to the Nulliporae ('the organisms with no pores'), a term derived from the great French naturalist Jean-Baptiste Lamarck. We know these now as the coralline algae, sometimes termed 'living rocks', which are forms of red algae that secrete so much calcium carbonate within their tissues that they become hard and rock like. Darwin could see that these formed a protective exterior rim to the reef, just a few tens of metres across and growing up above the low-water level, which took the full force of the ocean waves, and acted as breakwater to the complex structure within.

Darwin intuited that this reef atoll was a living and growing island, and explored, as far as he could, how it functioned. The main kinds of coral that he could see were the large colonies that grow near the sea's edge, such as the round masses of the genus *Porites*, which he had vaulted across with his 'leaping-pole', and *Millepora*.

How did the living corals relate to what the rest of the island is made of? Darwin studied the fish that lived on the reef, having been told by a local resident, a Mr Liesk, who was 'intimately acquainted with every part of this reef', that shoals of fish of the genus *Scarus*—that is, the parrot fishes—subsist by feeding on living corals. Darwin caught and opened up several of these fish, and found their guts full of coral debris, much of which was ground down to sand- and mud-sized particles. From among his extensive correspondence, he learnt from one 'Dr. J. Allen, from Forres' that holothurians, or sea cucumbers, 'which swarm on every part of these coral-reefs . . . subsist on living coral'.

The picture that Darwin pieced together from such fragments of information was one of skeletons of corals and other organisms, rapidly growing and just as rapidly converted into the calcareous sediment that accumulated on the beaches and on the sea floor in and around the atoll. Thus transformed, they were to form a large part of the structure of the island. He could form an idea of the nature and extent of this sediment from the soundings 'taken with great care by Capt. Fitzroy himself'. This remarkable, and ultimately tragic, captain of the *Beagle* used a weight on the end of a long rope made of lead, fashioned in the shape of a bell with a concave base, into which some tallow wax had been put. When this weight landed on the sea floor, it either took an imprint of the surface, if it was rocky, or caught some grains of sediment if the sea floor was soft. It was an ingenious and effective device, and allowed Darwin, in his mind's eye, to build up an impression of the island both above and below sea level.

Such production of skeleton-derived sands and muds has been amply borne out by later work. Not all of the 'facts' that Darwin built into his conceptual model of reefs have turned out to be correct. The good Dr Allen was mistaken. The sea cucumbers, for instance, are in the main sediment feeders, and do not feed directly on coral colonies—and so they are an important secondary reprocessor, and not a direct produ-cer, of the atoll sediments. But the important role of parrot fish, and other coral-eaters such as the notorious crown-of-thorns starfish, has been amply borne out by later work. Some parrot fish, with their unique 'beak', eat both the coral organisms and the underlying skeleton, and an individual fish can process as much as 5 tons of sediment through its gut each year. Others just nip away at the fleshy coral zooids, and do not touch the skeleton. Yet others feed on and remove the green algae (seaweed) that would otherwise smother the corals, thus effectively cleaning and maintaining the coral framework. Even where patches of bare coral skeleton are produced, these provide good landing grounds and attachment points where the planktonic coral larvae can settle, to form new colonies. Darwin would have been entranced by the subtle and delicate biological mechanisms by which the entire structure is main-tained. But in his work on coral reefs, it was the geological side of his enquiries that generally came to the fore. He was not just considering how these huge structures functioned at the present day; he mulled on how they built up through the vastness of deep time.

The Deeper Reef

Darwin marvelled, as did others in his time, at the contrast between the huge scale of the reefs and the 'soft and almost gelatinous bodies' of the 'apparently insignificant creatures' that built them. He also knew that the reef-building corals were confined to shallow, sunlit waters, while the coral-built structures seemed to persist to much greater depths in the ocean—which seemed entirely paradoxical. Not all coral reefs posed such difficulties. With shallowly founded reefs that fringed a landmass, he saw no difficulty—these just grew on any bedrock that was shallow enough to support the reef-building corals. But with coral atolls such as the Keeling/Cocos islands, there was a problem. Around these, the soundings made by Captain Fitzroy showed no trace of bedrock, but coral-derived sand that sloped off into deep water—and in places very steeply. In one place, only 2200 yards (just over 2000 metres) out from the breakers at the edge of the reef, the sounding line was reeled out to 7200 feet (nearly 2200 metres) depth, without finding bottom. That, noted Darwin, was a slope steeper than on any volcano, and he surmised steep submarine cliffs around the atoll's foundations at depth. So how can one build a coral island that extends kilometres in height above the sea floor, when the island-building process, centred on coral skeleton formation, only occurs in shallow water?

The answer is blessedly simple, and was signalled right at the beginning of Darwin's memoir. Coral formation started when what is now the base of the coral island was in shallow water. The island subsided steadily into deep water—but the corals continually grew their skeletons upwards so as to stay in the warm, sunlit water where they thrive. Where the island was originally a volcano emerging out of the water, a fringing reef could first form, and then build upwards as the volcano subsided. Eventually, when the whole volcano was submerged, the coral would keep growing upwards as the ring-shaped reef structure of an atoll. Darwin illustrated, from real examples, all the steps in this progression: from an island volcano with fringing reef, to an atoll with just the tip of a crater poking out in the middle of the ring-shaped lagoon, to just the atoll itself, with the volcano (so surmised Darwin, as he had no way of proving this) at depth somewhere beneath. With its simplicity and effectiveness, backed up by detailed evidence in the 214-page memoir, published in 1842, the idea quickly took hold. Charles Lyell—then Britain's leading geologist, and in many ways Darwin's mentor—danced for delight on reading the work.

It was not an uncontested hypothesis, though. The oceanographer John Murray, a leading member of the pioneering *Challenger* expedition between 1872 and 1876, suggested that atolls grew on top of banks of calcareous sand that had built up on the sea floor. The argument took a strange turn when Alexander Agassiz, the son of the eminent Louis Agassiz, entered the fray. Agassiz the father resisted Darwin's theory of evolution when it was published in 1859 and remained a creationist up to the end of his life; by extension he also resisted much of Darwin's other work—including that on atolls. His son, Alexander, became persuaded of the validity of Darwin's evolution theory—but nevertheless spent decades gathering and publishing evidence against Darwin's atoll hypothesis and in support of Murray's model. Attempts to resolve the question by drilling in the late 19th century were inconclusive—the boreholes were deemed not to have gone deep enough to provide clinching evidence.

Final proof would only come many years later, when a scientific team drilled through Eniwetok atoll in the early 1950s. The US military establishment, who were about to destroy this atoll with atom bombs, was persuaded to fund some deep boreholes simply to test Darwin's then century-old hypothesis. It resulted in its triumphant vindication. Two boreholes drilled through more than 1.2 kilometres of shallow-water reef limestone that, at its base, was over 30 million years old, to rest on dark basaltic rocks of an old volcano. It fitted Darwin's idea like a glove.

This is a *lot* of subsidence—more than Darwin had reckoned with. The mechanism that could produce this subsidence—widely enough in the oceans to explain the observed distribution of atolls—was only understood much later, with the advent of plate tectonics theory. Plate tectonics not only accounts for the slow drifting of continents around the globe, but also explains—and indeed predicts—vertical motions of the ocean floor. Thus, where ocean crust is forming by the injection of magma at mid-ocean ridges, such as the Mid-Atlantic Ridge, it is hot, expanded, and of relatively low density. Hence it floats high on the Earth's mantle, usually to within a kilometre or so of the sea surface; when heat and magma production is unusually high, the forming ocean crust can rise above sea level, as in the case of Iceland. As the ocean crust moves away from the mid-ocean crust, it cools and becomes more dense, and so contracts to descend to a lower level. Eventually, after several tens of millions of years, taking it hundreds or even a few thousands of kilometres from the ridge, it settles at an average of some 4 kilometres below sea level. Dotted on the surface of the ocean crust there are individual ocean island volcanoes, such as Hawaii, Ascension Island, the Canary Islands, and many others. It is these that, if built above sea level and if in the right climate setting, can acquire a fringe of coral reef. As the island sinks on being carried farther from the mid-ocean ridge on its steadily sinking conveyor belt of ocean crust, the fringing reef can take its skeletal growth upwards to form a ring-shaped coral atoll.

Not all fringing reefs make it so far. Scattered here and there in the oceans there are more or less deeply submerged, flat-topped mountains, called guyots. Sampling the rock at the top of these has yielded fragments of ancient corals, showing that these mountains, millions of years ago, used to be thriving coral atolls. Then, some vicissitude—perhaps a change of climate or sea level that the coral organisms could not adapt to in time—killed off the reef, and the rock production that goes with it. As we shall see, our own species threatens to be quite an effective producer of guyots in the near geological future. But first we may explore the wider setting, and the deep past, of today's reefs.

The Larger Frameworks

Coral reefs are the poster children of the shallow-water limestones that are forming today, and in many ways their skeletal frameworks are its defining feature. But the corals are only part of the story. Much of the extraordinary biodiversity of organisms they harbour have their own skeletons, which contribute to the resulting rock framework. Finding shelter among the coral thickets are bivalves, fish, crustaceans of different kinds, sponges, bryozoans, starfish, sea urchins, as well as those coralline algae—it is a menagerie of which the diversity and productivity are only equalled by the tropical rain forests and, unlike in the rain forest, where pretty much everything is recycled, here there is a progressive entombment of this array of skeletons where a reef grows upwards atop its sinking island foundations.

The reefs themselves, though, are often part of something larger. Gaze at the Bahamas islands from space, and their extent is not as great as that of the Great Barrier Reef. But what they lack in areal cover they make up for in bulk and longevity—and in their striking definition. It is not so much the islands that catch your eye, as a sharply defined, irregular patch of pale blue sea, over a hundred kilometres across, in which the scattered low islands—Andros Island, Grand Bahama, and hundreds of smaller ones—are embedded, the whole being set against the dark blue Atlantic Ocean waters that surround them. The greatest part of the pale blue patch is the very shallow waters—just a few metres deep—of the Great Bahamas Bank, a monstrous mound of limestone that has grown over 4 kilometres above the ocean floor that it rests on. It started growing in the Jurassic Period, some 150 million years ago, when this part of the ocean was near the infant mid-Atlantic Ridge, and close enough to sea level for corals and other organisms to take root and thrive, and it has carried on growing ever since, as that ocean floor has been carried over a thousand kilometres away from the active ridge, and more than 4 kilometres lower in elevation.

Yet only parts of this vast structure are coral reefs—mainly as a rim along the eastern edge, facing the breaking swell waves coming in from the Atlantic Ocean. The central part is a shallow lagoon, with few corals, but with a floor of lime mud which is in part just a chemical precipitate, but which in part is derived from another kind of skeleton—tiny crystals of calcium carbonate that occur within the tissues of some species of algae. When the algae die, the microscopic crystals are released as mud to the sea floor. This mud is home to other animals, particularly bivalves and gastropods which feed on the algae, and which add their own skeletons after death to the sediment. There are also areas of limestone sediment produced simply by chemical and physical processes, as on the western parts of the bank, where waves roll pinhead-sized grains called ooids along the sea floor; these areas of sea floor are too unstable and too poor in nutrition to support much in the way of a skeleton-forming animal community. These entire structures are called carbonate platforms—and sometimes semi-poetically referred to as 'carbonate factories' because of the efficiency with which they pull calcium and carbonate ions out of the seawater to build such enormous rocky constructions.

Though the coral reefs here are only part of the whole, they do much to define its shape, forming a tough and durable rim, which is not only a breakwater at the ocean surface, but also forms the edge of the kind of underwater steep slopes and cliffs that Captain Fitzroy's soundings picked out off the Keeling/Cocos islands, and that descend rapidly to the abyssal depths. These are mega-skeletons indeed, and they form a distinct part of the physiography of our planet.

Both the skeletons and the physiography were different, though, in the past. Huge as they are, they have also been sensitive to both the course of the biological evolution of the reef-builders, and to the course of planetary environmental change. Megaskeletons were often very different in the past.

Bivalve Reefs of the Cretaceous

The reef-building corals of today belong to a family called the scleractinian corals, which build their colonial skeletons of aragonite. The scleractinian corals range back through the current era, the Cenozoic, and through the preceding Mesozoic Era, when the dinosaurs lived. Over that time, they waxed and waned in their abundance and diversity, with three[53] major crashes, when reefs effectively disappeared as a major ecosystem from the Earth for a few million years.

One global reef collapse took place 55 million years ago, at the end of the Paleocene Epoch, when there was a natural outburst of some combination of methane and carbon dioxide from carbon stores in the ground into the ocean/atmosphere system. The result was a geologically brief, sharp global warming and ocean acidification event. The reef-forming corals were badly affected, with larger benthic foraminifera (like *Nummulites*), in rock-forming numbers (Figure 28), taking their place. Although the climate returned to normal after about a hundred thousand years, the coral reefs took much longer to recover.

Ten million years previously, another global 'reef gap' was associated with the mass extinction event coinciding with the asteroid impact at the Cretaceous–Paleogene boundary. The meteorite impact did not only decimate the coral reefs. It also wiped out another, more bizarre form of

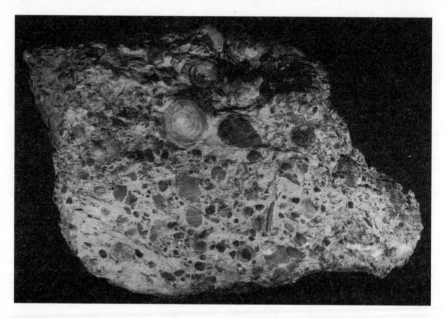

Figure 28. A rock slab of nummulitic limestone from the Paleogene of Spain. The slab is 14.5 cm long and near its centre there is a distinctive *Nummulites* with a spiral pattern.

reef that had arisen in the late Jurassic Period and developed mightily in the Cretaceous seas. These were built of a peculiar kind of bivalve shell, that of the rudist bivalves. In the classical rudist form, one part of the bivalve shell had evolved into a large, long conical form, and the other into a moveable lid. Both of these modified valves could take different forms, from bowl shaped to star shaped to an even more bizarre form, nearly 2 metres long, called *Titanosarcolites*, described as being like 'two giant, coil-toed Persian slippers placed heel-to-heel'. Given these extraordinary shapes, the rudists were long a puzzle, from their discovery and naming by Jean-Baptiste Lamarck in 1819 as large 'rude' or 'raw' objects common in the local strata. Lamarck did not know what they represented, and it was some time later that it was realized that these striking petrifactions, which can dominate limestone strata tens of metres thick, are a bizarre form of fossil bivalve mollusc.

The rudists grew in masses taking over large areas of sea floor, and particularly from mid-Cretaceous times onwards, began to displace the scleractinian corals. They came to extend over thousands of kilometres, being found from southern Europe to southern Asia and across north Africa—that, is lining both sides of the then-mighty Tethys Ocean, of which today's Mediterranean Sea is but a shrunken remnant—and also across the southern part of North America and northern part of South America (then separated by a wide seaway). This was the equatorial zone of Cretaceous times, when in general the Earth was hotter than it is now. It seems that the rudists adapted to these warmer, and probably saltier, waters rather better than did the scleractinian corals, and this helped them to expand into and dominate this enormous territory.

The rudist reefs must have been an extraordinary sight. The main conical shells, typically attached to the sea floor (and often to skeletons of their ancestors), grew more or less vertically, crowding together closely, not least to mutually support themselves in this position. The space between these upright cones was often more or less filled with sea floor sediment, so that all that would have been visible were the circular tops, looking like wide fields of tiny clustered craters.

As with present-day coral reefs, other organisms would live on and around this framework—such as corals (though here in a subordinate role), algae, sponges, and others. At the rims of the reefs, facing the open ocean, the rudist frameworks would be battered by incoming hurricanes, to be smashed into masses of shelly rubble cascading down the slopes that led to the ocean depths. The rudists proved resilient enough to regrow, time and again, following each such local catastrophe.

There is a legacy from this cycle of destruction and regrowth. The rudist-filled debris layers, being highly porous and permeable, were to become magnificent oil reservoirs, fully exploited by industrial-era humans. With this, the narrative threatens to come full circle: the living rudists pushed aside the reef corals in the Cretaceous, while now the carbon stores held within the buried cemeteries of their skeletons threaten, in driving global warming and ocean acidification, to inflict similar damage on the coral reefs of today.

The Ancient Reefs of the Palaeozoic

Both the scleractinian corals and the bivalve lineage that would eventually evolve into the rudists were beneficiaries of the greatest catastrophe yet inflicted upon the biosphere, in the volcanically driven, toxic world that developed at the end of the Permian Period. This mass extinction event—that saw the demise of the trilobites—drew down the curtain of the Palaeozoic world, which had its own distinctive assemblages of animals and plants, some of which combined to build up ancient reef structures.

The corals of the Palaeozoic, which succumbed to the end-Permian catastrophe, were structurally different to modern corals. There were two main kinds, the rugose and the tabulate corals. Many of the rugose corals were large, solitary, horn-shaped forms—somewhat similar in size and shape to the Cretaceous rudist bivalves described earlier, as a nice example of evolutionary convergence, while the tabulate forms—all colonial—formed smaller, interlinked arrays of tubes arranged in a variety of patterns; the skeletons of both groups were built of calcite rather than aragonite. Corals of these groups contributed to reef formation, but usually did not dominate reefs in the way that the scleractinians were to do later. Rather, the Palaeozoic reefs included sponges, coralline algae, and bryozoans, as well as corals, as inherent parts of the framework.

Reefs varied greatly in structure and strength throughout the approximately 300 million years of this Palaeozoic Era. At times, they formed robust, wave-resistant frameworks comparable to those we can still see today. Australia, for instance, hosts not just one, but two, Great Barrier Reefs. The less well-known one lies in Western Australia, in the Canning Basin. A basin in this case is a piece of the Earth's crust that has subsided over millions of years, in essence creating a long-lived, continually subsiding depression that, almost inevitably, becomes filled with sediments derived from the surrounding areas. The Canning Basin is a moderately large example, covering about half a million square kilometres, which subsided and filled with sediment for much of the early and mid-Palaeozoic Era. The ancient Great Barrier Reef is of mid-Devonian age, some 390 million years old, and its preserved parts stretch for nearly 350 kilometres, across the northern part of the basin. Unlike the modern Great Barrier Reef, one does not need a snorkel or glass-bottomed boat to appreciate the reef—all one does is walk across the arid landscape, where the rock exposures beautifully display the reef structures. There are no living reef creatures any more, of course, but the skeletal graveyard is magnificent.

This ancient reef contains a wealth of fossils, though these have been subject to the vagaries of underground transformation. Some are obscured by recrystallization. Others, though, have been accentuated by such subterranean processes. Some of the fossil skeletons, originally of calcite or aragonite, have been dissolved away, and the skeleton moulds have then been filled with silica. The silica casts stand out beautifully on the rock faces, where the sparse, though mildly acid rains have worn away the surrounding rock to leave the fossils standing proud (palaeontologists can help the process by dunking pieces of the rock in acid, so that these perfect, chemically transformed fossils can be extracted whole). The reef contains other, more practical, types of wealth too. Its cavity-rich, porous nature means that while it was buried deep underground it trapped fluids with hydrocarbons—some of which are still present at depth beneath the basin—and also fluids rich in dissolved metals, which converted it into a kind of underground Aladdin's cave, through filling some of those cavities with mineral ores of lead and zinc. There are many uses of a dead reef, and much of their 'academic' study has been carried out with a view to extracting the riches they contain.

The Canning reef is thought to have been a true, large-scale, robust barrier reef. But these kinds of structures did not persist throughout the Palaeozoic. The rise and fall of different groups of organisms through this time meant that the possibilities of building a wave-resistant framework fluctuated along with the fortunes of the main reef-builders. A major mass extinction event in late Devonian, for instance, decimated many of the main reef-builders of those times. During the succeeding Carboniferous Period, which is mostly known for its global spread of coal-forming forests, true reefs were rare. Skeletons then piled up in different ways.

The Carboniferous is known as a time of carbonate 'mounds' and 'buildups'. Like reefs, these are an accumulation of piles of skeletons of corals, brachiopods, sponges, and other organisms. But these did not generally form frameworks strong enough to resist the long-term pounding of ocean waves. Rather, they built up as mounds, generally starting their accumulation on sea floors that were below the reach of the waves. As generation on generation of these organisms lived and died and left their skeletons at the sea bed, the mound would build up towards sea level, with the additional mass being added by mud trapped between the skeletal debris, and by algae—seaweed—growing on them. As the mound built up towards the sea surface, waves would then break up and redistribute the debris.

This redistribution would make a difference to the Earth's solid surface, even on a planetary scale. Today, the coral reef structures, rimmed and strengthened by bulwarks of coralline algae, typically act to define the atolls and carbonate platforms, which then often plunge steeply—as Darwin and Captain Fitzroy were surprised to discover, on taking their soundings—into very deep water, with slopes often steeper than the angle of rest of piles of loose sediment. In the Carboniferous, generally lacking the stout reef frameworks, these kinds of large, steep platforms were rare. Instead of carbonate platforms, there were typically carbonate *ramps* at that time—long gentle slopes from a wave-affected beach reworking skeletal debris, gradually and steadily descending to greater depths. This is large-scale planetary physiography, controlled by the vagaries of organic evolution and extinction.

This kind of mass reworking of skeletons in shallow water was not restricted to the Palaeozoic. A classic example is the rock used to construct the pyramids of Egypt. This is a limestone dating back to the Eocene Epoch, just some 50 million years or so ago. It is made of foraminifera, shell-building protists—only these particular examples were gigantic, of the kind known as *Nummulites*, which can be over a centimetre across (these particular organisms, like all protists, were single celled, though the extremely large cells are thought to have possessed many nuclei). These foraminifer shells piled up on, and were washed across, the seabed in the Eocene as banks and sheet-like spreads, often deposited as 'event beds' during major storms. The resulting thick-bedded limestone was found to be perfect for making pyramids.

The oldest reefs constructed by multicellular organisms go back to the early Cambrian, some 520 million years ago. These first true reefs were constructed of an extinct, enigmatic group of organisms called the archaeocyathids. They are known just from their calcium carbonate skeletons, which typically comprised a couple of perforated cones, one placed inside another, with a root-like structure at the apex of the outer one that anchored the structure in the sea bed. This structure is a little like that of sponges, and indeed they may have been some kind of sponge. They flourished briefly—for little more than 10 million years— during which they built widespread reefs. Then, for reasons unknown, their populations collapsed in the mid-Cambrian, with a few stragglers persisting a little longer, but not surviving beyond the end of the Cambrian Period.

Prior to the Cambrian, the only structures that one might term reefs may have been accumulations of microbially bound sediment. How closely these approximated to true reef structures is questionable; certainly, they did not harbour anything like the biodiversity of the later reefs. And it is biodiversity that is foremost in our minds when we will later look not back, but ahead, to the uncertain future of the reef mega-skeletons of our own world.

The world of tiny skeletons, though, is the next realm to explore.

6

MINI-SKELETONS

To a Tudor physician covered head to foot in thick clothing doused in vinegar, and wearing a beaked mask, avoiding disease carried by airborne vapours was a serious business. Little was known about the vectors of disease, and little about how to cure it, and absolutely nothing was known about the microscopic world that was its cause. When Queen Elizabeth I of England was 'diagnosed' with smallpox in 1562, she was given the most advanced medical treatment known to the physicians of England at that time: being wrapped from head to foot in a red blanket. Apparently it worked, though no one could be sure why. The tiny virus that carries the smallpox disease is no more than 400 nanometres in maximum length, and there are one million of those nanometres in a single millimetre. The smallpox virus then, is much, *much* too small to see with the naked human eye.

What then, is the smallest object that the naked human eye can resolve? In good natural light, the eye can see objects somewhat less than 1 mm in size, and probably about one-tenth that, which is around 100 000 nanometres. Human vision can struggle to see the breadth of a single strand of hair. That is nowhere near small enough to resolve the smallpox virus. But it does mean that long ago, people had already recognized tiny, millimetre-long objects swimming in their drinking water. The indigenous Mogollon people of New Mexico, for example, noticed the minuscule ostracods (seed shrimps) that inhabited their lakes and streams, and depicted them 'swimming' on their pottery 1000 years ago. To see beyond the breadth of a hair, into a world of unseen skeletons, humans had to wait another 600 years, for the invention of the microscope.

A World Unseen

Magnifying glasses and eyepieces have been used since classical times, and what is perhaps more surprising is that no one thought to put two of them side-by-side, and make the first spectacles, until the 13th century. In optics, the early advances were made by both Italian and Dutch engineering wizardry. Spectacles were being used south of the Alps in the 14th century, and they had probably been invented in northern Italy towards the end of the previous century. Tommaso da Modena's painting of Cardinal Hugh de Saint-Cher in 1352 is the earliest depiction of someone reading and writing with the aid of spectacles. By the late 16th century, the Dutch spectacle makers Zacharias and Hans Jansen had gone one step further, and put several lenses in an extendable tube to make a microscope. It seems that people were using these devices to explore the world of both the very small and the very distant: at about the same time in Florence, Galileo Galilei was experimenting both with microscopes and telescopes. The Jansens' microscope could magnify perhaps nine times, while that of Galileo could manage 30 times. His microscope was elegant as well as functional. Conceived by Galileo, but engineered by Giussepe Campani, it was made of cardboard, wood, and leather, elaborately decorated, and inserted into an iron mount with three legs. But the great genius was too concerned with tracing the motions of celestial bodies to work out the exquisite detail of the insects he could see through his new magnifying device. It was his friend Johannes Faber who coined the name 'microscope'. On to the stage now enters Dutchman Antonie Von Leeuwenhoek, and, in England, Robert Hooke—one of the founding fathers of the Royal Society.

Antonie von Leeuwenhoek's contribution to microscopy is immense. From humble beginnings—his father was a basket maker and his mother

from a family of brewers—he is the first person in history to systematically document the microscopic world. His interest developed from
cloth, or rather a need to interrogate the integrity of that cloth using
better lenses. And he developed an improved technique for making these,
using tiny glass spheres that gave his microscopes the capacity to magnify 275 times—perhaps even more—representing a technological leap
forward akin to the invention of the scanning electron microscope in the
20th century. Leeuwenhoek was the first to observe bacteria, and the first
to observe myriad mobile single-celled organisms. His findings were so
revolutionary that for a time the Royal Society in London doubted him,
dispatching a group of 'wise men' to visit him in Delft, only to confirm
his findings. Leeuwenhoek lived a long life, dying at the age of 90, but
never giving away the secret of his microscope design.

At the same time that Leeuwenhoek was observing microorganisms in
Holland, Englishman Robert Hooke was working on his book *Micrographia*, and labouring with a compound microscope—one using several
lenses—that was probably inferior to that of his Dutch contemporary.
Hooke, too, was a man of humble beginnings, the son of a Church of
England curate on the Isle of Wight in southern England. He was a true
polymath, with an eclectic array of skills and friends—which included
Sir Christopher Wren, who built St Paul's cathedral in London—but he
infamously argued with Isaac Newton over who had conceived the idea
of gravity. If Newton won that battle, Hooke's impact on science is
nevertheless huge, from 'Hooke's law of elasticity' mastered by all high
school students, through sundry other discoveries that are testament to
his multifaceted genius.

*Micrographia, or some Physiological Descriptions of Minute Bodies made by
Magnifying Glasses with Observations and Inquiries Thereupon* was published in
1665. It was a revelation. For the first time in history, the general reader
could study the exoskeleton of a flea, a spider, and an ant, or gaze into the
eye of a fly. To keep his tiny subjects still while he drew them, Hooke
even resorted to the influence of alcohol, plying his ant with brandy to
keep it still for an hour. Published just before the Great Fire of London in

1666, the book also found its way into the diary of Samuel Pepys who, completely absorbed by it until the early hours of the morning, wrote that it is 'the most ingenious book that ever I read'.

The 'Ur Animals'

The thread of scientific discovery now passes to Germany, and the invention of the word that is used to describe these many tiny organisms, the 'Urthiere', being made from 'ur' primitive, and 'thiere' animals. For some 150 years after his correspondence with the Royal Society, the tiny organisms observed by Leeuwenhoek, and referred by him as 'very many little animalcules', were thought to be minuscule but fully formed animals. Leeuwenhoek himself had described one of these animalcules as being a tiny oval-shaped form that had two little legs poking out near the head. In 1765, German anatomist Heinrich August Wrisberg coined the term 'Infusoria' for these organisms, and as late as 1838, the distinguished German naturalist Christian Gottfried Ehrenberg still considered them as tiny animals, publishing that year *Die Infusionsthierchen als vollkommene Organismen* (which roughly translated means 'The infusion-animals as complete organisms'). At about the same time, though, Frenchman Félix Dujardin had already observed that some of these organisms were made of a single cell.

In 1818, 'urthiere' mutated into its Greek form 'protozoa'—a name meaning exactly the same, and one that has stuck fast in science down to the present day. Though German scientist Georg August Goldfuss coined the name, it was his compatriot, Carl Theodor Ernst von Siebold, who established its use for single-celled organisms. Siebold, incidentally, was also the scientist who established the single grouping for those animals with an exoskeleton and jointed limbs—the arthropods. The term 'protozoa' is now used mostly informally for a range of unicellular eukaryotic (i.e. nucleus-bearing) organisms that show functions that mimic animals: that is, they can move, and they can prey on each other.

In the early 19th century, one of these groups of protozoans, the foraminifera, was about to take centre stage in both oceanography and geology. Their fossils, abundant in ancient mudstones and limestone accumulating over millions of years, could be used to recognize different time intervals in rocks, whilst in the modern oceans their patterns of distribution could be used to map water masses of different salinity, temperature, and depth. Foraminifera possess one key feature. For more than 500 million years they have been making hard exoskeletons: a shell that protects their body from the elements. Their skeleton is sometimes almost literally homemade, stuck together from bits of sediment. Mostly, though, the foraminifer skeleton is formed from calcium carbonate, which the organism extracts from seawater. These shells have been preserved in their countless billions in rocks over hundreds of millions of years, reflecting, as time passed, the evolution of thousands of different species, which are characteristic of different ages of rock. Foraminifera are close relatives of the amoeba too, and so they are sometimes called the 'amoebae with a shell'.

An Amoeba within a Shell

Hooke and Leeuwenhoek established the foundations for the study of microscopic organisms. The mass manufacture of microscopes in the 19th century then brought that world much more widely within the reach of 'parlour microscopists'. During that century too, the biological relationships of many of the microorganisms that make tiny skeletons began to be recognized for the first time.

Born in 1802 and brought up along the Atlantic coast of France, Alcide d'Orbigny would, from an early age, be seen with his brother Charles collecting the local beach sands. From these he would extract the shells of many tiny organisms, including foraminifera; d'Orbigny became fascinated with these minute spiral forms that had long been classified as tiny cephalopods, relatives of the much larger and extinct ammonites.

As a young man, d'Orbigny set out to examine what these tiny shells might be, travelling first to Paris to study the collections of the famous naturalist Jean-Baptiste Lamarck. In 1826, d'Orbigny published a major study, *Tableau méthodique de la classe des Céphalopodes*, in which he coined the name foraminifera, which literally means 'hole-bearer'. By studying their minute forms—he eventually described over 690 different species—he had recognized that foraminifera have holes in their shells, and thus they were significantly different from their supposed larger cousins, the cephalopods, though at that time he still classified them within that molluscan group. The holes are indeed very important to foraminifera, because it is through these openings that thin strands of cellular material stream out from the shell to make pseudopodia (sometimes called 'false legs'), structures that foraminifera use to reel in food particles from their surrounding environment. It would be another decade though, before the true affinities of foraminifera were recognized, as unicellular protozoans.

Foraminifera have made a bewildering range of skeleton shapes that sometimes defy their classification as 'micro'. The shells of the largest of these, the *Nummulites* that lived over 50 million years ago, sometimes reach 16 centimetres in diameter, or about 300 times larger than the typical foraminifera. Compared with the size of an average domestic cat, a scaled-up *Nummulites*-sized cat would loom 100 metres into the air above you. This ability to change the size and shape of the shell identifies one of the key features of foraminifera, in that they are almost infinitely adaptable to a great range of aquatic environments on Earth. *Nummulites* adapted rapidly to fill vacant marine ecological niches following a geological catastrophe on Earth 55 million years ago, one that, for a time, exterminated coral reef systems worldwide.

Foraminifera are amongst the ultimate survivors. The earliest forms may have lived in the Precambrian seas more than 550 million years ago, and certainly from the early Cambrian onwards their tiny shells can be found fossilized in rocks. They live in a great range of aquatic ecosystems today, from freshwater lakes to the very deepest parts of the ocean, including the Marianas Trench of the West Pacific, where they thrive many kilometres below the sea surface. The foraminifer shell is a marvel of engineering too, even more so when you consider that it is made by an organism comprising just a single cell. The form of these shells has kept evolving through time. The marvellous *World Foraminifera Database*—yes, there is such a one—records 38 151 species of foraminifera as of January 2017, 8981 of them living, the rest being extinct as fossils.

The simplest of their skeletal designs are single-chambered sacs, tubes, or globular structures, while in some the tubes just keep growing, spiralling around to mimic the shape of a tiny, coiled serpent. These skeletons may be made of a resistant organic material, and sometimes sediment grains are embedded into this. Others stick sediment grains together to make a 'crazy paving'-style shell, as in the foraminifer *Reophax*.

As time went on, foraminifera evolved the ability to make their shells from calcium carbonate, extracting it from seawater to make thousands of different shapes. Some of the most ancient of these are called the fusulinids, which made multichambered shells that could resemble rice grains in shape. They became so common in marine sediments of the Carboniferous and Permian periods that they are widely used to make precise time zonations for rocks of that age.

The most diverse of living foraminifera are a group called the rotalinids. They are in the sands that Alcide D'Orbigny collected on the French coast: in typical sands that one can use to build sandcastles on a summer's day. The rotalinids make multichambered shells from calcium carbonate too, some types living on the sea bottom and some floating high in the water. They add chambers to their shell by stretching out their pseudopodia to make a protective cyst within which the new shell is formed. There are thousands of living sea-bottom rotalinid species, but only a few species—perhaps 40—that permanently float as plankton. Many of the floating forms have shells that look like tiny balloons stuck together, the most famous of these being *Globigerina* (Figure 29), the skeletons of which amass to form oozes on the deep ocean floor. Although the number of these planktonic species is very small, they are

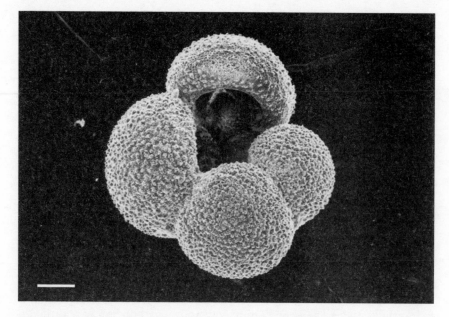

Figure 29. The planktonic foraminifera *Globigerina bulloides*. The scale bar is one-tenth of a millimetre.

all highly prized by scientists. This is because certain of them live in the cold, almost freezing surface waters of the polar oceans, others in the warm waters of the tropics, and yet others in the temperate zones in between. Change in these patterns of ocean temperature over geological time has largely been worked out by scientists studying the distribution of these fossils in ancient strata. To do this, though, their modern patterns had first to be discerned.

Challenging Times

The 19th century was an age of great discovery for biology, led by explorer-scientists such as Humboldt, Darwin, and Wallace. And towards the end of that century, building on the work of these giants, the Royal Society in England equipped the *HMS Challenger* to undertake the world's first global oceanographic survey. *Challenger* was not the first ship to survey the seas. During the American Civil War of the 1860s, Union forces conducted surveys down the east coast of North America. And, during that decade too, dredges from 270 fathoms by the USS *Corwin* in the Florida straits revealed a profusion of life that would soon help overturn the prevailing idea of a dead 'azoic' deep ocean.[54] In 1867, Scotsman Charles Wyville Thomson on the Royal Navy's HMS *Lightning* turned up ocean life in dredges from more than 600 fathoms off the Faroe Islands. HMS *Challenger* was therefore not the first to explore the deep oceans, but she was certainly among the most intrepid, and was the first to make a truly global survey.

Challenger was not originally equipped for science. Built in 1858, her 17 guns immediately identified that she was a warship, a steam- and sail-powered corvette in Britain's Royal Navy. She had seen active military service, being involved in operations in Mexico and the Pacific during the 1860s. But, with 15 of her 17 guns removed, the space could be converted into purpose-built scientific laboratories, one of these being for natural history. Sadly, little of *Challenger*'s structure remains. She was sold for scrap in 1921 and broken up for her copper bottom. But her figurehead was preserved, and it welcomes visitors to Britain's National Oceanography Centre in Southampton. *Challenger*'s scientific legacy remains as a wealth of materials from her 69 000-nautical-mile, 4-year journey that began in 1872. In the process, *Challenger* used both shallow and deep sea dredges to collect biological specimens from the oceans. Often these dredges were cast so deep into the sea, kilometres below the surface, that they returned to the surface crumpled, damaged by the immense pressures at these depths. But they mostly returned with a wealth of marine life, and amongst that life were myriad tiny creatures with hard skeletons. Two brothers were now set to describe these tiny creatures in enormous detail.

Born in the northeast of England at about the same time that foraminifera were first being recognized as protozoa, George Stewardson Brady (born 1832) and Henry Bowman Brady (born 1835) are two of the great names of early micropalaeontology, though almost entirely forgotten to most of science today. Still, they lend their name to a medal—The Brady Medal—presented once a year on behalf of the Micropalaeontological Society in London to honour an esteemed micropalaeontologist. The medal is lovely: cast in bronze and sculpted by famous artist Anthony Stones—a past president of the Royal Society of British Sculptors—it depicts the profiles of the two brothers on one side and, fittingly, a microscope on the other.

The Bradys were not by formal training natural historians. Henry was trained as a pharmaceutical chemist, whilst George followed his father into medicine. By the 1870s, though, when the chief scientist on HMS *Challenger*, Charles Wyville Thomson, was looking for someone to describe the microfauna, it was Henry who received the foraminifera, and George who took the tiny arthropods.

In science, ideas and observations are sometimes 'cooked' quickly whilst others simmer slowly. The Bradys were simmerers. Working with microscopes that would look primitive today, they set about systematically documenting the *Challenger* microfauna, producing beautifully detailed drawings of the foraminifera and ostracods that had been scooped up from the oceans. They produced two huge published volumes that proved to be scientifically durable, and that still grace the shelves of scientists today. The Brady brothers had provided the first detailed inventory of the microfauna of the oceans, from many different localities around the world. This formed the basis for recognizing that different water masses could be characterized by the organisms that lived in them—and that, on dying, fell to the sea floor, to become entombed in the accumulating strata. This is an idea now used widely by oceanographers as they study changes in ocean circulation.

'The uninitiated may be excused for wondering why men of ability should spend a considerable part of their lives in studying creatures so insignificant in size and so generally harmless to mankind, as the Entomostraca', as George's obituary in the proceedings of the Royal Society for 1922 recorded. It went on to say 'That it may be observed that, as in [the] old Camden's phrase, "many a little makes a mickle" and as little grains of sand may make a mountain, so the stupendous multitudes in which some of the entomostracan species occur make them indirectly yet ultimately important contributors to human food and comfort'. These words are a fitting epitaph for any scientist who studies small objects, from microfossils to fundamental particles. But they have a particular resonance now, as we begin to understand the tremendous importance of these tiny organisms so beautifully and precisely recorded by George and Henry. They occur in numbers in the oceans sufficient to underpin some of the major biological cycles on Earth, such as those that control the availability of oxygen, or the carbon cycle.

The tiny copepods studied by George Brady are perhaps the most abundant microplankton today. Typically just a few millimetres long, their arthropod bodies resemble those of the ostracods (Figure 30). But their exoskeleton is not toughened with calcite, and so after death their bodies rot, leaving almost no trace in the fossil record. They feed on phytoplankton in the surface waters, whilst their sea-bottom dwelling relatives mostly chew through organic detritus at the seabed, though some have parasitic or predatory lifestyles. In the Southern Ocean around

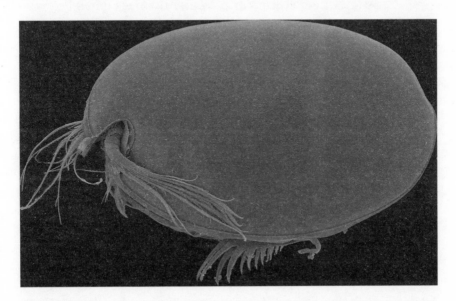

Figure 30. The ostracod *Vargula hilgendorfii*, from Tateyama, Japan. Specimen is 2.1 mm long.

Antarctica, copepods have become so super-abundant that, like krill, they are a primary food source for animals living in these cold polar waters.

The planktonic foraminifera live in waters that range from polar to tropical. Some of these species are 'sunbathers', favouring the warm surface waters of the tropics, like the elaborately named *Globigerinoides sacculifer*. Others favour cold waters, like the unpronounceable *Neogloboquadrina pachyderma*—its second name an allusion to having a shell surface that looks like an elephant's (pachyderm) skin. Still others chase the seasonal blooms of plankton in the mid-latitudes, like *Globigerina bulloides*. After death, their remains rain down to the seabed, to contribute to the slowly growing strata of the deep ocean, forming a permanent archive of ocean conditions at the time they were living. As time passed, over tens, hundreds, thousands, and ultimately millions of years, the patterns of ocean currents and climate changed above them, the changes being faithfully recorded in these tiny skeletons.

Ancient Mysteries and Whirling Dervishes

How far back does this archive of micro-skeletons go? Across the world in rocks older than the Cambrian System, older than 541 million years ago, it is difficult to find fossils. This seeming absence of fossils puzzled Darwin, who wrote 'Consequently, if my theory be true, it is indisputable that before the lowest [fossil-bearing] strata, long periods elapsed, as long as, or probably longer than, the whole interval from the [Cambrian] to the present day.' In these Precambrian rocks, organic fossils visible to the naked eye were later recognized: so too, eventually, were skeletal fossils quite invisible to the naked eye. But Darwin, even had he suspected their presence, was in no position to undertake the difficult and dangerous process necessary to release them from the rocks in which they are entombed. To extract these long-entombed miniature skeletons, one must use hydrofluoric acid, while dressed head to foot in protective clothing, rather like the Tudor physicians of long ago. Like smallpox or the plague, hydrofluoric acid is not to be toyed with. Although not the very strongest of acids,[55] it is highly toxic if swallowed or spilled onto skin, and causes severe burns. When the acid has done its work on the rock, though, the tiny fossils are released and can be examined by microscope. What, then, can be seen?

Ranging in size from a few tens of microns to a tenth of a millimetre in diameter, they have a variety of shapes. Most of the very ancient forms look like simple spheres made out of some complex, tough organic substance—so tough that it will resist the acid while the rock dissolves around it. The general term for them is acritarchs, coined by the palae-ontologist William Evitt in 1963, a name that means, in effect, 'we don't know what these organisms are'. We now know they go back an awfully long way into the Earth's deep past. A few years ago, convincing examples were dissolved out of South African rocks all of 3.2 billion years old.[56] These were just simple spheres (later ones could develop different shapes, or spine-like projections), but some of these pioneers were giants as such microfossils go—up to a third of a millimetre across. Their affinity, as with later acritarchs, was unknown. But they meant that, way back then, there were organisms, almost certainly single celled, that had learned to grow a kind of armour. What for?

Simple survival is the likely answer. Microscopic organisms today have, individually, a precarious existence. There is generally something out there that regards them as food (often another microbe), and so their lifespan is typically terribly brief. One way to circumvent this is to reproduce extremely quickly, so that many new individuals are continually recruited, to offset the carnage. But another way is simply to lie low at times when the going gets particularly tough. For some microscopic organisms that today live as microplankton, such as the dinoflagellates, this may involve literally growing a tough outer envelope, termed a cyst, and sinking to the sea floor. There, they wait, more or less inanimate, until better times come, when they emerge from their cysts, which are left discarded in the sea floor muds, and get back to whatever is normal life for them. We assume that at least some of the mysterious

Precambrian acritarchs, whatever they were, were doing something similar. Some have also been interpreted as egg cases, as the sheaths of microbial cells, and even as fungal spores.

When Darwin was writing *On the Origin of Species*, the tiny organic skeletons of acritarchs were unknown to him. minuscule organic-walled structures such as the pollen of modern flowering plants had already been observed in the 17th century using microscopy, though, and by the 19th century German botanist Heinrich Göppert was recognizing pollen fossils. By the end of that century the use of fossil plant pollen and spores to identify time zones in ancient rocks was becoming widespread, particularly in strata that were coal bearing, and these microfossils helped trace the evolution of plants.

Bill Evitt also correctly identified that some of the organic-walled microfossils in his samples were closely related to the cysts of dinoflagellates, protozoans that are abundant in the oceans today, with perhaps over 2000 different species. The fossils of dinoflagellates represent only their recalcitrant 'resting cysts' or dinocysts—perhaps analogous to the acritarchs—though only a fifth of the species of living dinoflagellates produce such structures. This might provide a clue to the obscure fossil record of dinoflagellates, as possible chemical traces of them have been detected in early Cambrian rocks from more than 500 million years ago, whilst the fossils themselves only appear in the Triassic Period, 300 million years later. This might mean that early dinoflagellates did not make cysts, and so could not be fossilized—or, alternatively, some of the enigmatic acritarchs might have been in reality dinoflagellates.

The dinoflagellates are extraordinary unicellular organisms, which, rather like microscopic versions of the scary protagonists of John Wyndham's novel *The Day of the Triffids*, have some characters of plants and some of animals. They bear two flagellae, one of which possesses a whip-like motion that both propels and rotates the cell like a whirling dervish, whilst the other flagella acts like a rudder. Hedging their bets, some use photosynthesis for their food supply, whilst others ingest food particles, displaying a range of complex feeding behaviour. Some engulf other cells using a flexible cell wall. Others are like tiny vampires, sucking away the contents of other cells via a straw-like device called the peduncle. Still others extrude a pseudopodia-like structure, the pallium, which flows around cells digesting their contents. The most famous relationship of dinoflagellates is their symbiosis with corals and many other organisms, where they live within the cells of their hosts, providing a food supply, and utilizing nutrients in return. This is a vital symbiosis, and when the dinoflagellates are expelled from corals, as happens during periods of intense environmental pressure, the corals 'bleach', losing their vivid colours and, starved of the dinoflagellates' food supply, eventually die.

Only some dinoflagellates form a protective outer skeleton in life. Some of those that do not make such an outer skeleton make an internal skeleton instead, this time of silica, with two star-shaped silica spicules located near the cell nucleus. These structures have been found as fossils too. Of those dinoflagellates that have an external skeleton, this is embedded within the cell covering as a series of overlapping structures made of cellulose, which is the stuff of cotton and paper. The skeletons that make up the resting cysts of dinoflagellates, which typically find their way into the fossil record, are made of a different material, a complex biopolymer called dinosporin that has some similarities to the resistant material in the walls of spores and pollen. The cysts may also be mineralized with calcium carbonate or silica.

Dinocysts are formed as part of the sexual cycle of the dinoflagellate (some acritarch cysts may have been similarly formed), and their structure can reflect changes in the water's nutrient supply, salinity, and temperature. The cysts are elaborate structures, with various projections and a kind of a trapdoor that is used by the organism when it finally emerges. This emergence can be as much as a century later (though most sojourns are much briefer) and so the cysts are as much like the suspended animation capsules of science-fiction films as they are skeletons.

Plankton in Glasshouses

In the original *Star Trek* episode 'The Devil in the Dark', Captain James T. Kirk and the crew of the *Starship Enterprise* encountered a strange silicon-based life form on the planet Janus VI that could burrow through rock. The science-fiction writer Isaac Asimov, similarly, invented animals he called 'siliconies', based on the way that silicon, like carbon, can form long-chain compounds—but not quite as well as carbon, so the siliconies were rather feeble, if charming, interstellar creatures. On Earth, silica *is* used by life forms—but not to form whole organisms based on exotic chemistry. It is remarkably widespread as a skeleton-building material, and may be the oldest medium from which the skeletons of animals are made.

Silica, usually in the form of quartz, makes up most of the sand grains on a typical beach, is a common component of many rock types, and is the main component of glass. It has the chemical formula SiO_2, or one atom of silicon combined with two atoms of oxygen. It is very slightly soluble in water, and is present in the Earth's oceans as silicic acid (H_4SiO_4), being bought in by rivers carrying materials weathered from the land, by submarine volcanic activity, and by dissolving pre-existing silica at the seafloor. So, although the idea of a skeleton made from a brittle substance like glass may seem a little strange, silica is a very widespread and readily bioavailable material.

The silica found in the skeletons of sponges and microplankton is not the simple form, as is present in crystals of quartz. Rather, it is biogenic (or opaline) silica, which is hydrated and amorphous, and similar to that in the gemstone form called opal, which is given its treasured opalescence by the water present in the mineral structure. Opaline silica is likely to be present on other planets too, having been recognized on the surface of Mars, near to volcanic rocks, and being regarded there as an indicator of hydrothermal conditions—that is, a mineral product of the chemical interaction between water and hot reactive rocks.

On Earth, the earliest group of organisms to use silica to make their skeletons may have been a motile group of protozoa called the choano-flagellates. They are considered to be the nearest living relatives of animals, with strong similarities to the flagella-bearing cells of sponges, a relationship already noticed in the 1840s by Félix Dujardin, a largely self-taught French biologist who nevertheless became a Professor at the University of Toulouse, and who contributed much to the understanding of protozoans. Alas, choanoflagellates have no fossil record, but this particular trail of evolution that led to sponges may have developed during the late Precambrian.

It is another group of protozoans that provides a longer-lived record of silica skeletons: the radiolarians (Figure 31). They are cousins of those shell-bearing amoebae, the foraminifera, and like them are widespread

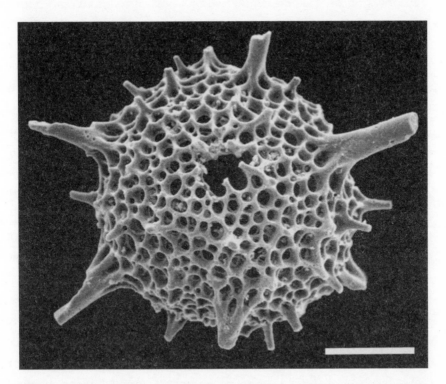

Figure 31. The radiolarian *Inanihella sagena* from the Silurian of Herefordshire. Scale bar is 100 microns.

microorganisms in the oceans. All radiolarians are planktonic; they are not fussy eaters, their food supply including other protozoa and tiny zooplankton. Like foraminifera, too, they can stream their cell material as 'axopods' to protect their delicate skeletons, catch prey, and dispose of waste materials. But unlike foraminifera, they form their skeletons from hydrated silica, perhaps from within the sheath of cell material that coats them. Why would such a small organism spend energy to build such an elaborate skeleton? A clue to this lies in the complex structure of radiolarians, the single cell of which is differentiated into inner and outer parts that control different functions. The inner part, the central capsule with the organelles and nucleus, is rather like the command centre of the cell. The outer froth-like envelope, which also contains the cell material that streams out to make the axopods and rhizopods—the same kind of structure as the pseudopodia of foraminifera—both makes this tiny animal buoyant and helps it catch prey. In turn, the outer skeleton of the radiolarian covers all of these tissues, and it often has a further protective layer of spines. Internally too, there are further layers of skeleton, with bars and beams extending inwards towards the centre of the structure that support the complex functions of the cell.

These elaborate skeletons—which often appear like tiny glass chandeliers—are hence amongst the most exquisite structures produced in nature, and a brief scan of the Internet will reveal that, minuscule as they are, they have inspired many artists, including the architects of the Skejby Hospital in Aarhus, Denmark. They inspired German scientist Ernst Haeckel too, who was the first scientist to bring the beauty of their skeletons to a broader audience through the 35 copper plates of his monumental work *Die Radiolarien*, published in 1862. Imbued with a love of both science and the arts, through his dual inspirations Alexander von Humboldt and Johann Wolfgang von Goethe, Haeckel is one of the key figures of 19th-century biology, bequeathing to us such terms as 'ecology' and 'phyla', and widespread concepts from school biology classes such as 'ontogeny recapitulates phylogeny' and the 'phylogenetic tree of life'. But like some 19th-century thinkers, he had a darker side too, believing in a hierarchy of humans, with 'higher' and 'lower' races. Little of this, luckily, spills over into the splendid illustrations of life that filled the pages of his *Kunstformen der Natur* ('Nature's Works of Art').

A little like the young Japanese scientist we met in these pages, Gengo Tanaka, searching for his bioluminescent ostracods at the side of the harbour at the Aitsu Marine Station in the summer of 2016, Ernst Haeckel's love of radiolarians also began at the seaside, on a visit to Messina in Sicily in 1859. The harbour there contained an abundance of radiolarians in the surface waters, and Haeckel became captivated by their intricate forms. Better still, they satisfied his two great passions, science and art. Little wonder then that *HMS Challenger*'s radiolarians were delivered to him to describe.

Most radiolarians live in the surface waters of the oceans, and many contain symbiotic algae—including dinoflagellates—that are photosynthetic. But some are found kilometres below the sea surface, as in the Marianas Trench. Like other plankton, they are important in the chemical cycles of the oceans, most notably that of silica. As they die, their skeletons sink into the deep ocean, to be preserved as siliceous sediments at the seabed. They therefore help to maintain a balance between the silica entering the oceans and that leaving. Until the emergence of diatoms in the Jurassic Period, radiolarians were the most important component of this cycle. The silica cycle, too, is linked to other ocean cycles, particularly that of carbon, as both radiolarians and diatoms are involved in the export of organic material from the surface waters to the deep oceans.

Yet other plankton form their skeletons from silica, but these are photosynthesizing forms that take us into the realm of tiny plants. Chief amongst these for their sheer numerical abundance and importance for aquatic food webs are diatoms. Diatoms are algae. They are a highly diverse group with perhaps 100 000 living species, more than all of the different types of vertebrate skeletons put together. In the oceans they are so numerous that they are probably responsible for nearly half the primary production—the basal food supply; this makes them, with the coccolithophore alga, the foundations of the entire ocean ecosystem. Some are adapted to freshwater, and they can be abundant in lakes and rivers. The diatom skeleton is made of hydrated silica, and is called the frustule. The silica of the frustule is made within the cell and extruded into the exoskeleton. Diatoms are typically very tiny, sometimes just a few microns in size, though some species reach gigantic—for them—proportions of a millimetre or more. Although there are a multitude of different shapes, diatoms come in two basic designs, elongate 'pennate' forms, and those that are circular, a little like microscopic hatboxes.

Diatoms compete with other skeleton-bearing algal plankton in the oceans for space. Where food is scarce they are outcompeted by the coccolithophores. But, they have supplanted the radiolarians as the key component of the silica cycle and there is evidence too that radiolarian skeletons may have become more delicate and less robust since the Cretaceous, as a result of increased competition for oceanic dissolved silica.

Perhaps the most enigmatic silica skeletons of all, and certainly the rarest, are those of the silicoflagellates. In contrast to the thousands of species of radiolarians and diatoms, silicoflagellates make up just a few species in the oceans and their opaline skeleton is internal, providing a framework to support the cell. Though they are also a part of the silica cycle, their delicate skeletons are much less abundantly preserved in siliceous sediments, forming perhaps 2% of this material. Propelled along by their single flagellum, silicoflagellates can photosynthesize like other algae, but strangely they may also form pseudopodia, perhaps to extract nutrients from seawater. Although there are only a few species of silicoflagellate, their skeletons can take many shapes, from tuning forks to triangles, making them like the 'snowflakes' of the ocean. Like other skeletal-bearing plankton, their different morphologies reflect changes in the ocean waters in which they live, and hence they too are skeletal archives of past oceanographic and climate change.

The Coccospheres

Sometimes the surface of the ocean turns a milky white colour from blooms of plankton that make their skeletons from calcium carbonate. Such blooms are seen in the late spring of the Celtic Sea between Britain and Ireland, for instance, or in the western approaches to Cornwall. They are large enough to be seen from space, and the plankton within them may reach millions of cells per litre of seawater. The most important of these carbonate-producing planktonic organisms, in terms of the amount of carbonate they generate each year, are the coccolithophores, tiny algae that characteristically bear two flagellae, and a third structure called the haptonema, which resembles a flagellum but is coiled. These are amongst the tiniest of the skeleton-constructing microorganisms on Earth, though they compensate in the length of their names, not least in that coccolithophores make an exoskeleton called a coccosphere (Figure 32), and each individual plate of that coccosphere is called a coccolith.

Although the whole coccosphere is typically just a few tens of microns in diameter, their numbers can add up to something much larger. During the warm climate conditions of the late Cretaceous, when there were no large polar ice sheets and sea level was much higher, much of the world's continental shelves were submerged beneath a deep sea. In these warm seas coccolithophores proliferated in numbers so great that their skeletons now make up the chalk cliffs of southern England and many other places. If you take some of this chalk rock and crush it between your fingers, you can then mix this with water and smear the mixture onto a glass slide. With a microscope you will see (just—for they are very tiny, and an electron microscope is needed to see them in detail) that it is largely made of coccolith plates, with some foraminifera and other calcareous fossils such as ostracods. It is this ability to bloom in such numbers, and their wide distribution in the oceans, that make coccolithophores so important for the life-support systems of Earth.

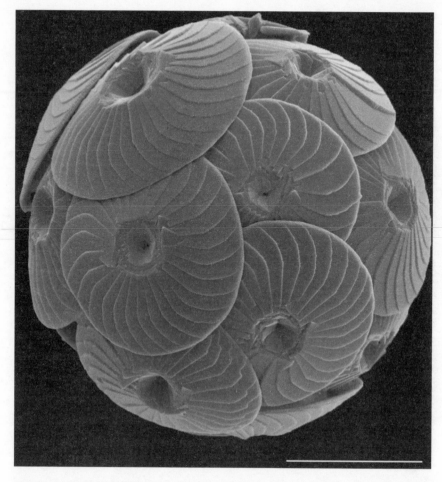

Figure 32. A coccosphere of *Calcidiscus leptoporus* subsp. *quadriperforatus* from a plankton sample collected in the Mauretania upwelling, Atlantic Ocean. Scale bar is 5 microns.

The skeletons of coccolithophores were noticed for the first time in the Cretaceous chalk rocks of Rugen Island, in northern Germany, by Christian Gottfried Ehrenberg in 1836. Ehrenberg was one of the pioneers of the microscopic world—and one who lived dangerously. As a young man he had taken part in an expedition to Egypt and the Middle East funded by Friedrich Wilhelm III and led by the charismatic but dubious Heinrich von Minutoli. It was a disaster, with fever killing several at the outset, and Minutoli soon abandoned the party to pursue his own individual travels. Ehrenberg was the only scientist to survive the gruelling 5-year journey, and was so disheartened[57] at the end that he left for others the study of the 80 000 specimens that had been collected.

His journey through the world of the very small, though, bore greater fruit. As a student he had discovered that fungi were not the spontaneous products of rotting vegetation, as was then thought, but instead reproduced via spores.

Initially, Ehrenberg thought these strange discs of the chalk to be inorganic, but later linked them to the 'animalcules' that he saw teeming in the Mediterranean, and realized that their remains, accumulating on the sea floor over immense spans of time, could eventually build up a layer of rock a kilometre thick.

Coccoliths were later collected during the *Challenger* expedition, and recognized to be the skeletons of microscopic algae. Each individual coccolith plate is produced from within the algal cell and then stuck on to the coccosphere from its inner surface. Why such a tiny photosynthetic alga should produce such an elaborate skeleton is unknown, but each coccolith resembles the shape of a Spartan shield, and when locked together in the exoskeleton they form a minute but formidable shield wall that must confer some protection to the cell.

Coccolithophores may be tiny, but their impact on the Earth is colossal. The oldest coccolith fossils are found in sedimentary rocks of the uppermost part of the Triassic Period, from a little over 200 million years ago. In the Palaeozoic oceans there were, perhaps with a few exceptions like planktonic ostracods, no oceanic microplankton that made their skeletons from calcium carbonate. This major pathway in the global carbon cycle became established midway through the Mesozoic Era, with the emergence of both the coccolithophores and the planktonic foraminifera as major players in the biosphere. Why is this skeleton transport system so important for life on Earth?

The export of these skeletons from the sea surface to the deep ocean is part of an ocean-wide process called the 'biological pump'. It occurs through the life and death of organisms but it is in fact a chemical pump, one that moves carbon in various forms from the sea surface to its depths, and back to the surface again. The carbon in ocean waters exists in organic form in the bodies of organisms, within their shells, and is dissolved as inorganic carbon in seawater itself, while the carbon accumulating at the seabed comprises the bodies of dead organisms and their shells. When a coccolith makes its skeleton near the sea surface it extracts dissolved carbon from the water and as the sea surface interacts with the atmosphere above, it modifies the content of carbon dioxide in the air, drawing down some of that carbon to maintain an equilibrium between ocean and atmosphere. After death, the coccosphere dissociates and the coccoliths (and their carbon) are dissolved back into the seawater or sink to the seabed. Those that become preserved in sea bottom sediments help to balance the amount of carbon that goes back into seawater and ultimately the amount in the atmosphere.

It is a delicate balancing act. Sometimes—as at present with humans burning fossil fuels—the amount of carbon released to the atmosphere exceeds the amount that can be quickly captured by the skeletons of the calcifying plankton, or that can be stored in their organic bodies. When this happens, atmospheric carbon dioxide increases, and so too does the uptake of carbon dioxide into seawater, which reacts to produce carbonic acid, with the subsequent release of hydrogen ions. That makes seawater less alkaline, and it is this process that is called 'ocean acidification'. Even then, the oceans do not become acidic—otherwise, they would rapidly become lifeless—because the carbonate that has accumulated at the seabed from generations of calcifying organisms simply dissolves back into the water to maintain the pH balance. This is one of many balancing acts that sustain the life of the oceans over very long time frames.

As well as helping to maintain a seawater chemistry that is just right for life, coccolithophores are probably responsible for generating about half of the ocean's primary food supply. They make their 'edible' tissues during photosynthesis, and a by-product of that activity is that they yield oxygen to the atmosphere. They are therefore—together with the diatoms—an integral part of the Earth's respiratory system. These skeleton-bearing microorganisms are not just a major component of the biosphere, but are also essential for maintaining its life support systems.

From the time of the Tudor physician to the present, our understanding of the world of the very small has grown enormously. The invention of the microscope, at the time that plague was still ravaging much of Europe, identified a new world of skeleton-making organisms, and the oceanic voyages of the 19th century showed their importance across a range of different Earth processes. These tiny skeletons and those of macroscopic animals like bivalves conceal more secrets that, once properly decoded, provide a long and detailed archive of oceanographic and climate change on planet Earth. That record provides us with vital clues that—if acted upon—can help humans mitigate the effects of future climate change. We will examine this record—after, though, exploring the world of aerial skeletons.

FLYING SKELETONS

The problem of overcoming gravity and achieving sustained forward movement in the atmosphere for objects that are heavier than air is one that has vexed humans for centuries, and certainly for much longer than the apocryphal apple that bounced off Newton's head. Human solutions have involved kites, spinning blades, buoyancy, and wings, all being types of construction that evolved successfully long ago, in prehuman nature. This is because the value of flight to organisms is clear—even to those with big skeletons. It allows you and your kin to disperse quickly, to find new places to live and new supplies of food. It helps escape predators—at least ground-based ones—and gives the ability to avoid seasonal extremes of aridity or cold that could threaten survival. Organisms that fly can benefit from some or all of these considerable advantages.

But lifting a skeleton off the ground has no simple engineering solution, particularly if that flight is to be sustained and directed. The elegant, long-eared African caracal, for example, just a little larger than a domestic cat, has learned to leap three times its length vertically into the air, and catch small birds returning to their nests. But this is a short airborne trajectory that requires just powerful hind legs and the mobility of a flexible spine. The caracal cannot fly. Its heavy bones are designed for the ground just like those of humans. Lifting a skeleton into the air for sustained flight needs a completely different anatomical design, as human pioneers of flight quickly appreciated.

The earliest of human flights lie in the realms of mythology, tragedy, and sometimes stupidity, but had already taken place long before the Montgolfier brothers sent people ballooning through the skies above Paris in October 1783. Records of kite-lifted humans are known in China from more than a thousand years before this, though this was not travel for business or pleasure, but for execution. There is a story of a famous early flight, too, that was taken by the Japanese 'Robin Hood', Ishikawa Goemon. He, like his English counterpart, is a semi-legendary figure who lived in the late 16th century, and was reputed to have possessed almost superhuman skills, not with a bow and arrow, but as a ninja. Goemon, with the assistance of a large kite, was said to have lifted himself to the top of Nagoya Castle in central Japan, from where he stole the scales of the two golden carp that adorned the top of its elaborate roof. These scales would have made for a profitable flight, for they had been hammered into shape from gold coins.

Kite-flying humans, like Ishikawa Goemon, were using perhaps the oldest technique for flight—one with its roots in the early history of Earth's biosphere—where lift can be sustained by the inherent energy present in the atmosphere, the energy that generates the winds.

Flying Kites

The air above is literally teeming with life, so much so that those Tudor physicians who believed that disease was carried by ill vapours may not have been so far wrong. The mass of tiny suspended organisms in the atmosphere is christened the aeroplankton, the study of which was started in the mid-19th century by Christian Gottfried Ehrenberg. Aeroplankton includes the spores and pollen of fungi and plants, bacteria, viruses, and a number of small animals too, including arthropods and arthropod eggs. Spiders, for example, will sometimes hitch a lift as part of this aeroplankton, unfurling their fine silk threads to catch the wind or rising air masses, as tiny kites. They may travel this way for just a few metres or sometimes for many kilometres. Whether or not these airborne biological masses constitute an ecosystem—or are simply transients on their way to the next port of terrestrial call—remains a matter of scientific debate. But some of these bugs might actually be able to live, metabolize, and reproduce in clouds at lower altitudes, and some have been implicated in changing the weather.

Bacteria have been found as high as 70 kilometres in the atmosphere, constraining the absolute limit to which Earth's biosphere extends. One of these is *Pseudomonas syringae*, a species of thin-walled bacterium (microbiologists term this a 'Gram-negative' bacterium) that has the remarkable property of causing ice to nucleate around it high in the air, and these ice-coated bacteria then act as seeds for water droplets. This is thus a cloud-making bacterium. Some forms of this bacterium live on the ground, but not all of these have ice-nucleation properties, which suggests that the airborne form is adapted to its aerial environment, just as much as is an albatross or a swift.

The idea of an airborne ecosystem has also found favour in the search for extraterrestrial life. For if microbes can exist in the Earth's atmosphere, perhaps they might have survived in the atmosphere of Venus too, a planet that once upon a time may have been a water-rich world similar to that of the early Earth. On Earth, microbial life has existed for 4 billion years—and sometimes in the most inhospitable places for life—and perhaps the aeroplankton was an early component of the dispersal mechanism of life. These passive flight systems, though, are limited. Even the silken sails of spiders do not allow for much direction, and the organisms therefore fly at the whim of air currents. This can be a wasteful form of locomotion, with many individuals perishing en route.

The development of more complex means of airborne locomotion had first to wait for skeletons to emerge onto the launch pad of land, and that would develop during the Devonian Period.

Arthropods in the Air

Some insects do not possess wings. There are the flightless silverfish that scurry across the pages of ancient books, or the misunderstood earwigs that do not occupy ears, but instead prefer to live in damp crevices or vegetation. And there are island flies that have lost their wings, perhaps because it is better not to be blown out to sea. Whilst there are many thousands of these flightless insect species, it is much, much harder to enumerate the more than 1 million winged species of insects that are alive today. It is not just in species numbers that insects have become the masters of the air. They sometimes flourish in vast numbers as individual species, to devastate the landscape around them. There was the notorious Rocky Mountain locust that hung in dense clouds over the skies of the American Midwest in the 1870s, cutting a swathe through vegetation in their hundreds of billions—or perhaps trillions—being the greatest agglomeration of animals that had ever been observed by humans to that date. These locusts were not to have the last word. Many American territories enacted laws that required all able-bodied men to destroy the troublesome insects during hatching time. It worked only too well, and by the early 20th century the Rocky Mountain locust had become extinct, another of those species to pass into history at human hands.

There are also the long, beneficial seasonal migrations of insects that occur each year. Over southern Britain, for example, at an altitude of 150 metres, scientists have measured some 3.5 trillion insects, amounting to a total of some 3200 tons of biomass, migrating seasonally,[58] northwards in the spring and southwards in the autumn. These huge transfers of insects are important for sustaining food webs, for pollinating plants, and for controlling pests. They are the food supply for other flying skeletons too, the insectivorous birds and bats. And they provide many of the mechanisms by which plant skeletons can reproduce.

This insect supremacy of the air is an old inheritance. It is a legacy of the earliest animal colonists of the atmosphere that probably developed the first sustained and directed flight some 400 million years ago. Having been the most diverse group in the early Cambrian seas, and the first animal colonists of the land, it was perhaps only natural, or perhaps inevitable, that arthropods would make their way into the air first, though the fossil evidence of quite how they achieved this is scant. In general, the delicate bodies of insects do not preserve well in the sedimentary deposits of the land, and so their fossil record is sporadic, and often limited to sites of exceptional preservation. One such site is the Rhynie Chert of Aberdeenshire, northeast Scotland. In the Devonian Period, some 400 million years ago, the landscape of Rhynie was populated with plants, fungi, and small animals—arthropods—and the silica-rich waters that periodically spilled out from the local volcanic springs flooded the adjacent landscape, petrifying and entombing its organisms.

It was William Mackie, a native of Aberdeenshire and a physician who practiced in the town of Elgin, who discovered the deposits of the Rhynie Chert.[59] Mackie had a strong interest in geology, having read natural sciences for his first degree at Aberdeen University, and he was drawn to the Rhynie area because of its well-displayed geology. In 1912, he collected the cherts—at first as loose blocks—and made 'thin sections' of these: thin, near-transparent slivers that, looked at through a microscope, allow geologists to examine the fine detail of petrified structures. It was these thin slivers of rock, like the pages of an ancient book, that revealed a marvellous array of fossil organisms to Mackie. Many plants were found there, showing preservation even of cellular structure, and also arthropods—and these include *Rhyniognatha hirsti*, the earliest known insect in the fossil record.

You might be forgiven if at first you did not recognize *Rhyniognatha hirsti* as an insect. Only fragments of its head and mouthparts are preserved, almost ghost like within the chert.[60] But the morphology of its mandibles indicates a relationship with a group of winged insects called the Metapterygota, a group that includes beetles, bees, and dragonflies. Although wings have not been preserved on *Rhyniognatha hirsti*, its similarity to the Metapterygota suggests that it might have been able to fly and so, early in the Devonian, arthropods may already have found their way into the air. Only much later in the Carboniferous are fossils found of wings themselves, but absence from the intervening rock record does not necessarily mean they were not there. Flying insects are naturally delicate constructions, and most such fossils are only exceptionally preserved, in particular geological materials such as chert or amber.

When they are preserved, it is often the insect wings that are found as fossils. These wings are outgrowths of the arthropod exoskeleton, arising from the second and third segments of the thorax, the part of the body that lies behind the head. Like the arthropod skeleton itself, the wings are sclerotized—toughened—and consist of lower and upper layers that form a membrane separated by veins, some of which carry blood and nerve tissue, and tracheae for breathing. The wings are moved either by muscles attached to them directly—for example, in dragonflies—or by muscles that flex the thorax to which the wings are attached. In the latter group, which includes most of the winged insects, flexure of the thorax allows the insect to beat its wings much faster.

How did these intricate structures first develop? There are many different theories for this, but unfortunately almost no fossil evidence. However, lobe- or leaf-shaped structures borne near the base of the limbs of aquatic crustaceans may hint at the origin of wings. These structures, which help respiration and chemical regulation, can be identified on fossils of some of the earliest crustaceans from the Cambrian. They are called epipodites, and they seem to have the same origin as wings. But how would a structure used for breathing in an aquatic crustacean evolve into the lift mechanism of an airborne insect?

Perhaps these structures first evolved to be useful for aquatic 'flight' within water. At some point they may have been raised above the level of the water so that the arthropod could skim along the surface, a bit like an ancient windsurfer. Beating of these structures would then have allowed powered skimming, and it is just one small step, or leap, for an insect to take to the air on the first true wings.[61] These steps have not been discovered in the fossil record, and there is a big jump in time from

Rhyniognatha to the first fossil wings. The oldest insect wings to have been found fossilized are already well developed.

The greatest of all these wings are those of *Meganeura*, a giant Carboniferous insect related to dragonflies. *Meganeura* had a wingspan of 65 centimetres, much larger than that of the largest living insect, the damselfly *Megaloprepus caerulatus*, which reaches a wingspan of 'just' 19 centimetres. *Megaloprepus caerulatus* is a predator, turning the tables on spiders, which it feeds on, and the large size of *Meganeura* suggests that it too may have been a predator. But why was *Meganeura* so much bigger than its living relatives? One possible answer to this may lie in the development of Earth's first large-scale forests during the Devonian and Carboniferous periods. A by-product of building large plant skeletons is the oxygen released by photosynthesis, and this may have accumulated to levels in the Carboniferous atmosphere that were greater than those today, and so might have allowed the respiratory systems of arthropods to support a bigger body. It is a pattern that is also seen in living arthropods, in some (mostly) tiny marine crustaceans called amphipods, that grow to be largest in waters with more oxygen.[62] Carboniferous insects may also have got big simply because there were fewer airborne predators to consume them, for mammals, birds, and reptiles had yet to colonize the air.

From Pterosaurs to Terror Birds

Insects colonized the air not long after they had reached the land, and possibly as early as the Devonian, some 400 million years ago. Hence, they were some 170 million years ahead of the first flying reptiles—the pterosaurs—and about 250 million years ahead of birds and mammals in achieving powered flight. Their takeover of the aerial domain matched their early dominance in the oceans, and of the land. In this they were probably helped by their small size, and the hollow construction of their skeleton, as opposed to the heavier internal skeleton of vertebrates. Vertebrates may have needed tall structures to act as launch pads, and the development of forests may therefore have been an integral part of learning to glide, parachute, or fly.

The very first vertebrates to experience flying through the air, though, were not reptiles, birds, or mammals, but a fish. *Potanichthys* is a flying fish from the mid-Triassic that lived about 240 million years ago in the seas of what is now Guizhou, China. It was a small fish, just 15 centimetres long, but it possessed enlarged pectoral fins and a forked asymmetrical tail similar to those characteristic of living flying fish (Figure 33), and it is this shape that suggests it could have glided above the waves for several tens of metres,[63] enough time and distance to escape from predators, just like its modern counterparts.

Modern flying fish are sometimes reported to glide for up to 400 metres, but to sustain flight well beyond this distance requires powerful wings with muscles. The first vertebrates to use this mechanism for flight were the pterosaurs, successful reptilian flyers that navigated the skies for more than 150 million years from the late Triassic to the end of the

Figure 33. Flying fish.

Cretaceous Period. No dinosaur movie is truly complete without a pterosaur in the skies, though pterosaurs themselves are not dinosaurs. And in popular culture, even King Kong did battle with them in the 1933 movie that bears his name.

The description of the first pterosaur did not identify a flying animal at all (Figure 34). The fossil was recovered from a quarry in the Solnhofen Limestone, in rocks about 150 million years old, near the town of Eichstätt in Bavaria, southern Germany. In 1784 this pterosaur was described by the Italian Cosimo Alessandro Collini, curator of the Cabinet of Natural History Curiosities at the palace of Charles Theodore, the Elector of Bavaria. Collini had formerly held the position of secretary to the great 18th-century French philosopher Voltaire, only to lose this position when he apparently insulted the philosopher's long-time companion Marie Louise Mignot.

Collini's mistake—with the pterosaur and not with Madame Mignot—was to give greater credence to the fossils associated with the pterosaur—that included fish—rather than to the animal's outlandish morphology, and as a result he made the assumption that this strange animal too was aquatic. News of the discovery passed through French natural historian Jean Hermann to that most famous of 18th-century geologists, Baron Cuvier, and both came to the conclusion that the greatly elongated fourth finger of the animal supported a membrane that may have formed a wing[64]—though the membrane itself was not preserved in the specimen. Nevertheless, this fossil was a revelation about the potential antiquity of vertebrate flight, and it would soon be clear that the Jurassic skies were populated by many more such beasts.

Another fossilized pterosaur was found by that resolute carpenter's daughter, Mary Anning, who excavated it from the crumbling and perilous cliffs of Lyme Regis in 1828. Baron Cuvier, who had been impressed by her previous discoveries of ichthyosaurs and pterosaurs, was now well and truly amazed. Charles Dickens, who had been following these fossil discoveries, described them with gusto. Cuvier, he said, now gave the 'palm of strangeness' to this newly uncovered 'monster half

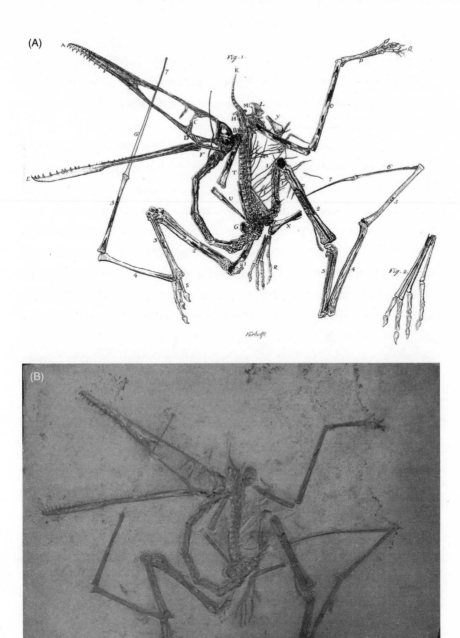

Figure 34. (A) Original illustration of *Pterodactylus antiquus* by Italian naturalist Cosimo A. Collini. (B) Photograph of the original rock slab with *Pterodactylus antiquus*. The rock slab is 23 cm by 16 cm.

vampire, half woodcock, with crocodile's teeth ... and scale armour'. Dickens himself thought them 'grewsome beasts', and considered them thoroughly unsuitable as pets—for any potential Jurassic pet owners who might have existed, that is.

The gruesomeness of their reputation was only enhanced when Arthur Conan Doyle featured them in *The Lost World*, with one pterosaur snatching the hero's roasted meal as they prepare to climb up to that world, and another flying to freedom at the end of the novel. In between they were described in such terms as 'a crawling flapping mass of obscene reptilian life' or 'like some devil in a medieval picture'. Conan Doyle's repulsion was almost tangible—yet, in reality, for those potential Jurassic pet owners, pterosaurs might have not been completely short of appeal. Most or all of them were covered in what have been called 'pycno-fibres'[65]—and so the pterosaurs were, in effect, furry.

They were undoubtedly spectacular. Prior to human flight, the pterosaurs were the largest objects ever to fly through the heavens. But which of these was the largest? Many pterosaurs are known only from fragments, and in such cases scientists estimate the overall size of the animal by scaling the individual bones that are preserved to those of other, more fully preserved fossil pterosaurs. Three types of pterosaur, all from the late Cretaceous, are genuinely monstrous. The Jordanian *Arambourgiania*, the Romanian *Hatzegopteryx*, and the American *Quetzalcoatlus* all had wingspans exceeding 10 metres.

The pterosaurs included giants, but there was much more to them than the monstrous size that some of them attained. Their skeletons are like those of no animal alive, and Cuvier's astonishment, and Dickens's glee, were fully justified. The body is squat, with a huge head—as regards braincase as well as the long beak—atop a squat chest and abdomen. The hind limbs are well developed, and the front ones simply extraordinary: there is a short upper arm bone, attached to a rather longer lower arm—but that makes up only one-third or so of the structure. The rest of it, to the tip, is mainly a single, enormously elongated fourth finger. It is a unique solution to the flying problem, this limb—finger, mostly—

supporting a wing that is a membrane of muscle and skin—covered with those pycnofibres. This membrane (pycnofibres, too) has sometimes been found fossilized, and in at least some pterosaurs it was also attached to the hind limbs, as a kind of whole-body wing.

Those bones, too, played the classic weight-against-strength game that all flying animals must play. That enormous skull had large holes in it, to cut down its weight. The bones as a whole developed extremely thin walls, especially in the later pterosaurs, and many were filled internally with air. But those bones still needed considerable strength to provide anchors to powerful muscles, and not to break with the first flap of those enormous wings. And so, thin walled as they were, they still needed many internal struts and buttresses. The microstructure of the bones was multilayered, involving spiral 'bone bandages', providing extra strength. The breast bone and shoulder bones, in particular, needed to be large—to provide attachment surfaces for those large flight muscles—and robust. Nevertheless, these could not be delicate, 'ultra-light' skeletons—and nor are those of modern birds. They have to be large and strong enough to withstand the forces involved in maintaining the animal in the air. The engineering properties of those skeletons became increasingly optimized through the Jurassic and Cretaceous, but they could not escape from physical laws.

How did they fly? Probably by a mixture of flapping and, once aloft, soaring, using thermal updrafts of air, as do modern vultures and condors. One traditional picture of pterosaurs sees them swooping down on wings to catch their prey, just as the hapless Raquel Welch, in the movie *One Million Years BC*, was plucked from the ground by a hungry pterosaur (fortunately for the plot she was dropped into the sea before serving as lunch). Skimming for prey from the surface of the sea was certainly used by some pterosaurs, and the *Rhamphorhynchus* of the Solnhofen Limestone are sometimes preserved with prey between their bills. Here too in the limestone are instances of predatory fish—*Aspidorhynchus*—attacking a pterosaur as it flew just above the water, a confrontation that resulted in death for both the fish and pterosaur, as the two became entangled, and the fish could neither separate itself from nor eat its cumbersome prey. Many pterosaurs probably only rarely used this airborne skimming technique as a way to feed, and their skeletons suggest they exploited a diversity of different feeding patterns, seeking out prey on the ground, and using their robust jaws to deliver a fatal knockout blow to small animals.

Pterosaurs have been found dating back to the late Triassic. But, despite many different species now being recognized, the origins of pterosaur flight remain hotly debated, largely because there are too few fossils that show intermediate characteristics that might link with possible flightless ancestors. However, from their skeletal morphology, the pterosaurs certainly belong within the archosaurs, the group that also includes crocodiles, dinosaurs, and birds. And some clues to the origins of pterosaur flight might be suggested by the anatomy of *Scleromochlus*, a tiny late-Triassic archosaur that reached just 18 centimetres long. *Scleromochlus* lived some 230 million years ago, its fossils being found in the rocks of the Lossiemouth Sandstone, near Elgin in northeast Scotland. It had long hind legs that may have enabled it to hop and spring great distances, and this style of locomotion might be seen as a precursor to flight. Such a hypothesis, though contentious, might also be suggested by interpretations of an upright stance for early pterosaurs that means that some were competent terrestrial walkers and runners.[66]

Though they were by far the largest flying animals of the Mesozoic, and indeed of all time, pterosaurs were not the only vertebrates in the skies of those times. They were joined by gliding squirrel-like mammals such as *Volaticotherium*, which lived in the forests of Asia, probably a little more than 160 million years ago during the Jurassic,[67] and by flying dinosaurs, such as the chicken-sized *Microraptor* of the early Cretaceous that possessed four—rather than two—feathered wings. And in the Jurassic skies there was another group of animals that provide a link with the most diverse of modern flying vertebrates. In the same rocks that yielded the first pterosaur to Collini—the Solnhofen Limestone—there is

Archaeopteryx, a kind of missing link between non-avian dinosaurs and the avian branch. *Archaeopteryx* possesses feathers, though it is not a bird in the strict sense, possessing jaws with sharp teeth, a long bony tail, and a hyperextendable second toe that relates it to feathered dinosaurs, such as the dromaeosaurids, which include *Microraptor*.

The first near-complete specimen of *Archaeopteryx* ever found (only 12 are known in total, all from the Solnhofen Limestone) is now the most valuable specimen in the possession of London's Natural History Museum. There is no precise estimate of its worth (there is little to compare it with) but 10 million pounds has been quoted for this petrified carcass, which was already considerably decayed, and missing most of its head and neck, by the time the unfortunate animal was buried in the limy oozes of the Solnhofen lagoon that it had fallen, or been swept, into. Nevertheless, the splayed and distorted bones that remain, with the feathers forming a fan off each wing and off the long, bony tail, make it a fossil as immediately recognizable and iconic as any in that venerable institution's impressive collection—or indeed any in the world. The museum originally needed some luck—and some fancy footwork on behalf of its superintendent, the politically astute Richard Owen—to acquire this marvel.

The specimen had been sold to a local doctor, Karl Häberlein, by the quarry owner, probably as payment for medical services rendered. The good Dr Häberlein—more astute than the quarry owner—knew that this fossil was special and was ready to haggle. Local attempts to purchase it for the Bavaria state collections were stymied as the key academic, Andreas Wagner, Munich's Professor of Palaeontology, had been keen to describe this new sensation but was too baffled by this 'mongrel' of bird and reptile, which to him should not exist, and (he was a creationist) too concerned it might incite Darwinian interpretations, to recommend purchase. Dr Häberlein then negotiated with the London museum, asking for £750. Owen had been told not to spend more than £500. Eventually a down payment of £450 was agreed, with £250 to be paid later. Dr Häberlein used the money to marry off one of his daughters.

That was surely useful, but the Natural History Museum likely got the better deal.

So the Mesozoic skies were busy with the flapping, soaring, and gliding of a range of winged animals, just one of which was *Archaeopteryx*. By early Cretaceous times, some 130 million years ago, pterosaurs were joined in the skies by the kinds of birds that have a recognizable direct connection with those living today. The earliest of these birds—*Archaeornithura*—is found in the rocks of Hebei province in northern China and it possessed modern bird features such as fan-shaped tail feathers and a U-shaped wishbone. It seems to have been an expert flyer too, and to have been specialized for the types of wading habitats that many birds adopt today.[68]

The avian branch of the dinosaurs survived the calamity of the end-Cretaceous mass extinction and went on to proliferate in the Cenozoic world. It is unclear which characteristics of this dinosaur group allowed them to survive, while many others went extinct, but it may have been the wider range of habitats they evolved into, their superior skeletal anatomy for flight, and perhaps also evidence in their skeletons—including in *Archaeornithura*—for faster growth to adulthood.

The Birds

The birds, as the surviving dinosaurs, have carved out an effective niche for themselves—one in which they are arguably more resilient than were their extinct, monstrous, non-avian kin. While they do not have a monopoly on the aerial realm, and are not restricted to it, they have certainly made themselves the most consistently visible parts of it across most of the world.

Like the pterosaurs, they have fashioned effective wings—but ones of fundamentally different design, in which the arm and wrist bones make up most of the structure, with the fused digits forming merely the short termination to the structure, instead of most of it. And of course the wings, and indeed the whole bodies of the birds, are clothed with feathers, not pycnofibres. The bones, like those of pterosaurs, have to tread that fine line between weight and strength, and so are light and hollow—and yet sturdy. The beak forms part of that equation, being lighter than a standard toothed jaw—and yet, through various morphological adaptations, being effective for a range of purposes, from the tearing of flesh to the crushing of seeds. Even among closely related species, they can be adapted in shape and structure to different functions—as Darwin famously saw among the 15 or so closely related finches of the Galapagos Islands, where the beaks—adapted to the local food source on each island—vary widely in size and shape.

Modern birds show great diversity, from the condors, hawks, and eagles, through parrots, crows, and ravens, to starlings, sparrows, and wrens, with many specialized types—the hummingbird that can hover over a flower to sip its nectar, say, or the swift that is peerless in the air but helpless on the ground, to the seagoing penguin. Compared to the pterosaurs, they show many small, highly agile species—those sparrows and others—but none now reach the sheer size of the largest pterosaurs. That is not to say that it is impossible for them to evolve to such a size. Back in the Miocene, in South America, there was *Argentavis*, a bird something like an enormous condor with a 7-metre wingspan—and so approaching, if not exceeding, many of the large pterosaurs. Perhaps the entire structure of the biosphere has now changed so much that such giants no longer have a viable place. There were, in truth, other giant birds, and in some abundance. But they mostly kept their feet firmly on the ground.

Back to the Land: the Terror Birds

While the Cretaceous–Paleogene calamity saw the end of the classic dinosaurs as popularly known, some of the niches of carnivorous dinosaurs were to be reborn among birds that came down to earth to become giants. Within a few million years the terror birds had emerged in South America to become formidable predators in their own environment.

The seriemas of South America are the only living close relatives of these terror birds, and, all being less than 80 centimetres in height, are now ferocious only to small amphibians, reptiles, and insects. But their mode of feeding gives a hint of what their extinct kin could achieve, flipping their hapless prey against hard objects to stupefy them and then break their bones, and then ripping away at the flesh with their sickle claws.

Their ancient relatives, more properly known as the phorusrhacids, were considerably more formidable, being armed with axe-shaped beaks and sharp, heavy talons and standing up to 3 metres tall. The terror birds lived in South America, where they thrived for most of the Cenozoic Era, until just a few million years ago. The skeletons are clearly those of predators, though there remains debate about whether they mainly dined on large or small prey. Their skull was built for powerful downward slashing movements, rather than the side-to-side shaking that could tear a victim apart (their claws may have served that purpose).

The decline and disappearance of the terror birds is also something of a mystery. Some have speculated that their disappearance was the result of the tectonic changes in Central America that finally lifted up the Isthmus of Panama to join the two Americas together, permitting the exchange of animals between the two continents, and so greater competition between predators. Yet some terror birds, like the 2.5-metre-tall *Titanis*, actually travelled north then, and for a while, thrived in North America. And, at about the same time that the terror birds evolved in South America, another group of large birds—some also standing 2 metres high—became prolific in North America, and they too were predators armed with formidable hooked beaks. But in contrast to the terror birds, they cohabited in their North American landscape with many large mammal predators such as the cat-like nimravids, and seem to have competed well with them, perhaps by targeting different hunting grounds and prey.

A human origin for the demise of these giant birds is also very unlikely, as most (perhaps all) of them were extinct long before people arrived in the American landscape. More likely, the loss of these giant birds—in both North and South America—reflects a subtle interplay of many factors, including climatic ones. And, after the Isthmus of Panama was formed, a reorganization of ocean currents caused a cooling global climate. The impact of this on habitat, perhaps linked to the interchange of animals between north and south, may have been a large factor in the disappearance of the giant birds.

Giant avian dinosaurs lived on well after the terror birds, often in geographically isolated settings away from large mammal predators, especially humans. The herbivorous moa survived in New Zealand until the arrival of humans in the 13th century. Some of these birds were truly gigantic, with one bird, *Dinornis robustus* reaching over 3.5 metres, taller than any terror bird. A faint echo of these giant birds also survives in the emu of Australia and the ostrich species of Africa. These are omnivores, but their great pace gives some inkling of the horror of being chased down by an emphatically carnivorous terror bird. For that in the end is the final epiphany for these awesome birds. Here in a chapter on flight, they could of course not fly. It is in the vast diversity of 'normal' flying birds, numbering in the thousands of species, that the legacy of the dinosaurs is really continued. They do, though, have some competition in the air, from the bats.

Flying Mammals

For much of this story of flying skeletons, mammals have been standing on the sidelines, not yet able to join the race. Three groups of animals have really dominated the skies: insects, pterosaurs, and birds. But there is also a long history of aerial colonization by mammals. In 2017, perhaps 4 billion human flights will be made,[69] and at an average weight of 62 kilograms per human, that means 248 million tonnes of human biomass will fly through the air. Or we could put this in another way and measure the weight of flying human skeletons, and that will be about 37 million tonnes (and that is not counting the giant exoskeletal planes that those humans will climb into to fly). Although this does not compare with the gigantic flying biomass of insects and birds, mammals are making up for lost time. There are, too, over a thousand species of that most successful of mammal flyers, the bat, and they are found in all terrestrial landscapes barring the high latitude Antarctic and Arctic.

The first airborne mammals had much more modest beginnings. Early flyers like the Jurassic *Volaticotherium* were gliders, using techniques similar to those of modern flying squirrels. *Volaticotherium* is known from only a single fossil in China, but this is highly instructive, and shows an animal adapted for an arboreal lifestyle with grasping toes. *Volaticotherium* also possessed a membrane extending between its four limbs and to the base of its tail, allowing it to glide between the trees of its Jurassic habitat. And its mouth housed long canine teeth suggestive of a carnivorous diet which, given its small size, probably meant it was an insect eater.[70]

All flying mammals bar two groups, bats and humans, are gliders like *Volaticotherium*, and this includes flying squirrels and possums. The first evidence for powered flight using wings is not recovered from the fossil record until the Eocene, perhaps more than 100 million years after the flights of *Volaticotherium*. These 'first' bats are already fully formed, and like the earliest history of insect and vertebrate (pterosaur) flight, there are few intermediates to fill in the gaps of the how and why of their flight. Nevertheless, one of these earliest bat fossils provides some telltale signatures of how these mammals first mastered the air.

Onychonycteris is a small bat preserved in the rocks of the Eocene Green River Formation of Wyoming, USA. This sedimentary deposit is famous amongst fossil hunters the world over for its fish, which are often on sale in fossil shops and on display in museums. The fossils of the Green River Formation also include birds, and all of these organisms—fish, birds, mammals—came to rest in the waters of what were a series of intermountain lakes some 50 million years ago.

Onychonycteris has features of its skeleton that are a little different than those of modern bats. Its hind limbs were longer and its forelimbs shorter, a resemblance to animals—like sloths—that spend much of their time hanging from trees. Its wings were short and broad, and it may not have been able to fly the long distances of some modern bats, perhaps preferring to hop between trees. These features suggest an origin for the flight of bats in forests, perhaps linked with gliding, the ancient technique adopted by *Volaticotherium*.

Bats have evolved wings that are a wonder of biomechanical engineering. It is little wonder that when Count Dracula wishes to transform himself from vampire to flying creature of the night, he uses the form of a bat rather than a bird or insect. The mechanics of bat flight depend on a thin membrane that stretches between the highly elongated digits of its forelimbs, and this elongation is already present in bat fossils from the Eocene. It is yet another model for a wing, akin to the pterosaurs in the elongation of the digits, but retaining all five of them, instead of simply relying on one. And because the bat's wing is literally alive, the membrane is equipped with a series of sensors that the bat can use to modify the shape of its wing as it flies, to make flight highly efficient and reduce drag. As a further adaptation to help their agility in flight, bat bones are remarkably flexible, as bats 'renew' them in life by replenishing them in collagen—a fact that has caught the attention of medical researchers who wonder whether there is an ability here that might one day be mimicked by our own osteoporosis-prone species.

Bat wings have adapted for a whole range of different ecologies. There are the superfast insect hunters, which have long and pointed wing tips, and can sometimes reach short bursts of horizontal flight approaching 100 mph,[71] the equivalent of running the 100-metre sprint in just over 2 seconds. The hoverers, seeking out stationary insect prey or seeking nectar from flowers, have short rounded wingtips that allow for maximum manoeuvrability.

In bats there is a group of mammals that probably developed an airborne lifestyle from living within the tree canopy, that hunt insects

Figure 35. Fruit bat.

and occasionally small birds, but that can also be vegetarian, like the fruit bats (Figure 35) or, like the vampire bats, haematophagous (a word that Count Dracula would have appreciated). Bats are themselves sometimes eaten by other birds, such as bat hawks and bat falcons. These different kinds of bat strategy probably evolved long ago, perhaps as long ago as the Eocene.

This marvellous range of complex flying skeletons has taken hundreds of millions of years to evolve. Working out the timing of all these events is part of the reconstruction of the long history of the Earth as a whole. Much of that broader history is written, in great detail, within the bones and shells of the organisms this planet has sheltered. And it is skeletons as witnesses to and recorders of planetary change that we turn to next.

SKELETON ARCHIVES

Methuselah is said to have lived for 969 years between about 6000 and 5000 years ago, and he was, according to ancient biblical texts, the longest-lived person in history. In these texts, he was said to have come from a family of very long-lived individuals, being the grandson of Adam and Eve, and the grandfather of Noah. Or perhaps scholarly translation, or mistranslation, might have been involved, as totting up his years simply as months gives an age of just over 78, a more typical human lifespan.

Humans now typically live to be into their ninth decade, and those who reach 100 years are no longer so unusual. Indeed, as we write, the Italian Emma Morano has just passed away, at the age of 117. She was the last person alive who was born in the 19th century, her birthday being November 29th 1899. Emma Morano lived through extraordinary changes. She was a small girl celebrating her fourth birthday when Orville Wright took to the air for the first powered flight on December 17th 1903. She was a teenager of 14 years old when the assassination of Archduke Franz Ferdinand on 28th June 1914 precipitated the First World War, and she was approaching middle age—by normal human standards—at 39 years old when the German invasion of Poland on September 1st 1939 began the Second World War. When Neil Armstrong took the first step on to the surface of the Moon on July 21st 1969 she was approaching old age, but that is already nearly half a century ago. As Emma Morano's mind took in all of these changes, her bones and teeth became an archive too, reflecting such things as the changing diet of Europeans over more than 100 years, the burning of fossil fuels by her fellow humans, and the explosion of atomic bombs by a few of them. One hundred and seventeen years is a long time from our personal perspectives, but to understand the profound changes to planet Earth over its more than 4-billion-year history, we have to dig much deeper and find archives of information—many in the skeletons of ancient organisms—that have preserved records over such a span of history.

Methuselah and Emma Morano notwithstanding, the skeletons of primates may not be the best archives of past environmental change. For a start, our nearest great ape cousins have similar lifespans to us: the bonobo, chimpanzee, gorilla, and orangutan all clock in at between 40 and 60 years. Some animal and plant species live for much longer. And, as they live, these organisms add to their skeletons, layer by layer, through life. In a good year, with a balmy summer and a rich harvest of food from the land or sea, their skeletons grow rapidly. In a bad year, when the food supply fails, growth may stall. For an animal like a coral or a bivalve mollusc, the changes to the environment around them might show as growth spurts or breaks in the skeleton. And if these growth breaks can be reliably matched between shells, there is the possibility to put together chains of years stretching back over centuries and millennia.

In taking such a journey backwards in time, beginning near the present, and finishing more than 550 million years ago, the nature of the evidence changes. At the beginning of the journey, some of the fossil organisms that one encounters still have living representatives. Later on, these will disappear. Further back still, the fossil archives have been buried, deformed by mountain-building processes, and their chemistry and mineralogy altered. And, in the Precambrian, there are few skeleton archives that can be used; it is difficult, though not entirely impossible, to glean such a picture of the past from such ancient rocks. One might begin, though, with a molluscan Methuselah that—had it not been for a lethal encounter with a shipboard freezer—would probably today have still been extending a quite remarkable lifespan.

A Mollusc for all Seasons

Arctica islandica is a mollusc that lives in the cold waters of the North Atlantic. It is colloquially known by many names, including ocean quahog, Icelandic cyprine, and 'black clam' (Figure 36). Ocean quahogs are edible, though they are said to have a very strong taste. They are harvested as shellfish along the east coast of North America. The animal grows quickly to sexual maturity, adding many increments to the shell after just a few years, the shell typically reaching about 5 centimetres in diameter. Thereafter, they grow much more slowly and some, over a very long time, reach a maximum size of about 12 centimetres. An ocean quahog collected in 2006, just to the west of the small Icelandic island of

Figure 36. *Arctica islandica*, a bivalve Methuselah from Iceland. Scale bar is in cm.

197

Grimsey, holds the record for the most long-lived animal ever recorded. It has been christened 'Ming',[72] after the Chinese dynasty founded in 1368. Ming, along with many of its kin on the seabed, was hauled aboard the Icelandic fishery research vessel *Bjarni Sæmundsson* to meet an unfortunate end, especially after its distinguished 507-year lifespan. Ming was frozen on board the ship, and thawed out for scientific investigation back onshore. When the shells of these ocean quahogs were examined it was quickly realized that several of the others had also lived for more than 300 years. Ming was by far the oldest, though, born in 1499, and hence about midway through the 276-year long Chinese dynasty. For what seems like an eternity to humans, Ming quietly lived on the seafloor of the North Atlantic. Ming was already 121 years old when the Pilgrim Fathers passed to the south on their way to the Americas. When the colonies of the United States of America declared their independence from Britain in 1776, Ming was approaching its 277th birthday. The age of Ming the mollusc can be measured by the number of growth increments on its shell, and by comparing those patterns with other ocean quahog specimens and looking for overlaps in the patterns of growth, it has proved possible to extend this skeleton archive back yet earlier, to the mid-7th century, to Dark Ages Britain under Anglo-Saxon rulers.

What stories do these venerable ocean quahog shells tell? One animal, collected from the waters of Iceland in 1868, had lain in the collections of the Zoological Museum at the University of Kiel, Germany for over 100 years, when Bernd Schöne, a palaeontologist at the Johannes Gutenberg University in Mainz, and his colleagues rediscovered it. This particular ocean quahog does not have a name, so one might here christen it Carl, after the founder of the museum in Kiel, Carl Möbius. Like Ming, Carl was also very old, and reckoned to be 374 at death. That takes Carl back to the later part of the 15th century, to about 1494, the year of birth of Suleiman the Magnificent, greatest of the Ottoman emperors. Schöne and his team set about analysing the shell of Carl in minute detail.

First, the team cleaned the shell so that they could examine and count its growth increments from 1494 to 1878. Next, they sampled the early part of the shell—when Carl was growing rapidly—drilling out carbonate from the skeleton to extract isotopes of oxygen and carbon that give information about sea temperature and food supply. Back in 1512, for example, when Carl was 18 years old, they could discern that this mollusc had lived in summer waters with a temperature of about 9 °C, while normally Carl lived in waters a little cooler, at about 6°C. A little over 300 years after Carl was born, telltale patterns in the shell showed slower growth during the years 1816 to 1818 that seem to reflect a reduced food supply, and perhaps also cooler surface waters then, in the North Atlantic.

In the Northern Hemisphere, 1816 is recorded as the 'year without a summer'. The year before, in April, on the tropical east Indonesian island of Sumbawa, the volcano Tambora erupted, blowing away its uppermost kilometre and ejecting tens of cubic kilometres of lava and ash into the air. The explosive eruption was so loud that it could be heard in Jakarta nearly 1300 kilometres to the west. The eruption column of ash reached up to the stratosphere along with large amounts of the gas sulphur dioxide. These materials were then blown around the globe by high-level winds, the sulphur dioxide forming sulphur aerosols that eventually fell back to Earth to be recorded in Greenland ice as a sulphate-rich ice layer. There were yet broader climatic effects. The global dimming caused by the pollution in the atmosphere blocked out the sun's light and made 1816 one of the coldest years in the past 500 years. In North America and Europe, crops failed and people starved. Joseph Mallord William Turner's paintings of that year show a red glow in the sky that may literally reflect particles from the eruption in the upper atmosphere over England, while Lord Byron, his spirits lowered by the cold and gloom, then wrote his bleakest poems—and became a muse, too, for monsters, as we shall see later. In the North Atlantic, Carl silently recorded this chill weather in the slower growth of its shell.

Why does this environmental archive deep in the history of Carl's skeleton matter to our story? It matters because when looked at collectively, the shells—exoskeletons—of these ocean quahogs from the North Atlantic provide a record stretching back perhaps 1400 years, long before humans made historical accounts of the weather or climate from this region. For their latitude, the waters of the North East Atlantic are warm, steeped in the heat of the Gulf of Mexico, from where these surface waters come, to be carried across the Atlantic by the great ocean conveyor, a system of currents that connects the ocean surface to its depths, and the deep Pacific with the Atlantic and Southern Oceans. Ocean quahogs may help to monitor the strength of this ocean conveyor system as it has changed over centuries in the North Atlantic, and hence help determine the controls on North Atlantic climate. Shells of this mollusc from Iceland show cooling waters in the 14th century that presage the interval of time in Europe known as the Little Ice Age, and this was perhaps caused by a weakening of the Gulf Stream. Again in the 17th century, their growth slowed, this time possibly associated with cooler seas and a reduction in sunlight called the Maunder sunspot minimum. This is the time when the Thames River in London regularly froze over in winter, so that frost fairs were held, uniting Londoners south and north of the river. The first of these fairs was in 1608, but the most famous one was during the severe winter of 1683–4, when the Thames upstream from (old) London Bridge was frozen over from late December to early February.

The ocean quahogs show yet another dip in growth during the early 19th century—bracketing the time of the last frost fair in London held in 1814—associated with the Dalton sunspot minimum, and later compounded by the eruption of Tambora.

These shells are eloquent about the sensitivity of the North Atlantic to climate change. The twin continental masses of North America to the west and Europe to the east help deflect and focus the warm waters of the Gulf Stream north-eastwards, feeding the air above with moisture that seeds the ice sheet of Greenland with snow. As these long-lived molluscs show, the climate here is subject to substantial and periodic changes between warmer and much colder spells. And as the fossil record is explored into deeper levels of time, this skeleton archive can be used to observe the Greenland ice sheet growing to cover much of the landmass of northern North America and Europe.

The Saw-tooth Foraminifera

The near-surface waters of the oceans often abound in a planktonic foraminifer called *Globigerina bulloides*, a species that was first described by Frenchman Alcide d'Orbigny in the early part of the 19th century. Although individuals of *Globigerina bulloides* have short lifespans, their antecedents have occupied the seas for several million years since the Pliocene Epoch. *Globigerina bulloides* builds its skeleton from calcium carbonate, growing a series of globular chambers, the last one being the largest. Long-dead specimens look like four interlocking balloons arranged spirally around a large central aperture, but in life its glass-like skeleton is covered in a protective mesh of spines. Though it is a small foraminifer, with a diameter of less than 1 millimetre, *Globigerina bulloides* is widely dispersed across the oceans in the zone that is penetrated by sunlight, and it is particularly abundant in regions rich in the phyto-plankton it feeds on.

Like the ocean quahog, skeletons of *Globigerina bulloides* record the properties of the ocean around them. These properties include saltiness and temperature, but the foraminifer skeleton can also sense, from a distance of thousands of kilometres, the amount of ice at the poles. This may seem absurdly far-fetched: for how can a planktonic foraminifer living in the near surface waters of the tropical Atlantic Ocean off the coast of East Africa sense what is happening in the high polar regions? Yet the foraminifera possess exactly such sensitivity. Fossils of *Globigerina bulloides* tell stories of ocean waters from millions of years ago.

To grasp how the skeleton of a long-dead marine creature can do this, the passage of oxygen through the water cycle on Earth must be followed—that is, we must track what happens to the O in H_2O, as that molecule moves across the planet. Oxygen is an abundant element, the third most abundant in the Universe. On Earth it makes up 21% of the atmosphere, 90% of the mass of the oceans, and just over half of the mass of the Earth's crust. Oxygen occurs in three different stable forms— isotopes—all born within different layers of a star late in its life. Oxygen-18, written ^{18}O, has ten neutrons and eight protons in its nucleus and is the heaviest of the isotopes, representing a little more than 0.2% of the oxygen in Earth's atmosphere. ^{17}O has nine neutrons and eight protons and accounts for a little less than 0.04% of that oxygen. ^{16}O is the commonest form, accounting for 99.76% of the oxygen in the atmosphere, and has the same number of neutrons and protons, there being eight of each.

The ratio of two of these isotopes, ^{18}O and ^{16}O, can be traced from seawater, through water vapour, rainfall, rivers, lakes, and ice. Because water with the lighter isotope ^{16}O more readily evaporates from the surface of the oceans, water vapour in clouds is relatively enriched in ^{16}O. And conversely, following that evaporation, the water left behind in seawater or a lake is enriched in ^{18}O. Water with the heavier ^{18}O also tends to fall out of clouds more easily as rainfall, which is logical given that it is heavier. This means that as clouds move inland and sequentially lose more rain, the water remaining within the clouds becomes progressively enriched in ^{16}O. As a result, snowfall at the centre of a large continent like Antarctica is enriched in ^{16}O. When the ice sheets of Greenland or the Antarctic grow, they become long-lived reservoirs of ^{16}O, whilst the oceans become relatively enriched in ^{18}O. And that enrichment is then locked into the skeletons of foraminifera, and bivalves, and many other carbonate-making skeletal organisms.

The realization that seawater variations in ^{16}O and ^{18}O are preserved in the skeletons of ancient organisms, and can track changes in climate over great lengths of time, was a discovery made in the late 1940s by Harold Urey at the University of Chicago. Urey is perhaps best known in science as one half of the Miller–Urey team that demonstrated how complex amino acids could be synthesized in the Earth's early atmosphere during electrical storms. But it was his work on the fractionation of oxygen isotopes (specifically ^{16}O and ^{18}O) between water and carbonate that would make his greatest impact on science, founding, in effect, the entire discipline of palaeoclimatology. It began with a paper in the journal *Science* in 1948 that reconstructed Cretaceous sea temperatures from the carbonate of fossilized skeletons of belemnites, torpedo-shaped cephalopods that lived in the Mesozoic seas.

Urey noted from the very beginning that in order to extract reliable ancient sea temperatures from fossil materials, a large and reliable dataset was needed. These fossils, too, must be perfectly preserved, as any change from their original composition would alter the primary chemical signal. In reality, much of the material from the fossil record has been changed, as skeletons are buried and subject to heating, pressure, and the migration of fluids through rock. The older the material is, the more problematic it becomes to interpret. Even so, it is possible to find pristine, and largely unchanged, fossil material even in very ancient rocks. Even where some change has occurred, a remnant environmental signal can often still be discerned.

Bivalves provide a very detailed record of climate—sometimes on a seasonal basis—over hundreds of years, and documenting changes over geological timescales requires a near-continuous record of the same or similar creatures making their skeletons for millions of years. And these skeletons need to lie undisturbed on the seabed on which they have fallen—and then to be buried by the 'eternal snowfall'[73] of countless subsequent generations of skeletons. Foraminifera are perfect for this, as they proliferate in huge numbers in the water column, and are preserved on a deep-sea floor that is undisturbed by the action of waves, tides, or currents. And there they accumulate, over millions of years, preserving a continuous record that can then be recovered by drilling through the ocean bed and recovering a core of sediment layers.

Although the oceanographic importance of foraminifera had been suspected at the time of the *Challenger* expedition in the 1870s, it was really through the work of three pioneer palaeoclimatologists, Jim Hays, John Imbrie, and Nick Shackleton,[74] that their importance for palaeoclimatology was realized. Their work followed over 100 years of calculations on how changes in the Earth's orbit might impact on climate, beginning in the mid-19th century with Frenchman Joseph Alphonse Adhémar. Later, Scotsman James Croll and Serbian Milutin Milanković would build on these ideas, recognizing three important orbital cycles: the eccentricity of the Earth's orbit around the sun, operating on a 100 000-year cycle; obliquity, the changing angle of the Earth's rotational axis, operating on a 40 000-year cycle; and precession, the wobble on its axis as the Earth spins, operating at about a 20 000-year cyclicity. Their calculations had shown that as a result of these orbital changes, the amount of sunlight reaching the two hemispheres of the planet would vary, and as a result this could be an important pacemaker of the growth and decline of ice sheets.

Hays, Imbrie, and Shackleton coupled their knowledge of these astronomical cycles with new geological data emerging from sediment cores produced as part of the Deep Sea Drilling Programme. They focused on fossils from two drilling sites in the southern part of the Indian Ocean between 40° and 50° latitude south. Their analyses were published in 1976 and showed, for the first time, physical evidence that the variations in the Earth's orbit had indeed exerted a control on major changes in the extent of northern hemisphere ice during the past half million years. This was clearly displayed by the oxygen isotope data they extracted from *Globigerina bulloides* skeletons which, for very many millennia, had been sinking from near-surface waters to accumulate in sediments of the southern Indian Ocean. When these data were plotted, they resembled an earthquake seismogram that seemed to increase its magnitude periodically. The jumps, towards more enriched ^{16}O signatures in the skeletons, are quite sudden on the graph, and represent rapid changes to warmer seas and less polar ice that seem to be paced by the 100 000-year eccentricity cycle, with smaller, but still significant, changes on 40 000-year and 20 000-year patterns. These major jumps were followed by long-term declines in sea temperature and steadily increasing ice volume as more ^{16}O was removed from seawater. Short, warm interglacial periods, then, were separated by long, complex, cold glacial periods—the 'ice ages' of vernacular speech. The overall pattern mapped out a 'saw-tooth' profile that has been replicated in many subsequent studies, analysing the skeletons of many different foraminifera.

The pattern of change over the past half million years in the northern polar region, recorded in the Indian Ocean foraminifera, is part of a much longer story of global change. Oxygen data from foraminifera at many ocean drilling sites around the world were steadily amassed to show a detailed record of climate, with gradual cooling from 55 million years ago, interspersed with intervals of more pronounced warming and cooling. There is a particularly big jump in the isotope values about 33.6 million years ago that marks a phase of rapid growth of the Antarctic ice sheet, whilst the gradient of the curve towards more ^{18}O-enriched oceans steepens through the past 3 million years, recording the development of ice sheets in the northern hemisphere.

The periodic switches between more ice and less ice identified by the foraminifer record reflect another impact of climate change, in that they document swings in sea level that often exceed 100 metres. For the high peaks of the Himalayas such a sea level rise makes little difference, but for low-lying areas, like the coastal plain of the Brahmaputra, it is profound. For an atoll in the Pacific Ocean just a few metres above the sea surface, the change is irrevocable in the short term, shifting the ecosystem from land to sea in just a few years. How life responds to such sea level fluctuations may be gleaned from another archive of skeletons—in this case through a story of Indian Ocean atoll tortoises—a story, here, that does not rely upon subtle chemical signals. What matters with these tortoises is simply the presence or absence of their quite unmistakable gigantic skeletons.

The Return of the Giant Tortoises

Aldabra is an isolated place—a small group of four islands that form an atoll in the Indian Ocean, lying some 600 kilometres to the west of the

East African coastline, and to the north of Madagascar. The atoll sits atop a giant submarine mountain and has a complex history that extends back to the Pleistocene Epoch, a history that shows the strong influence of changing sea level influenced by the 100 000-year-long cycle of the Earth's orbit around the Sun, as it stretched out from a circle to an ellipse and back again.

The youngest of three main limestone layers that make up the atoll is called the Aldabra Limestone. It formed on the floor of a very shallow sea about 125 000 years ago,[75] at the time of the last interglacial warm period when global sea level, at its peak, was probably a few metres higher than it is today. After that peak of warmth, climate cooled and sea level dropped until about 17 000 years ago, when polar ice had grown to its maximum level, drawing water out of the oceans so that sea level was about 120 metres lower than it is today. This was followed by a similar amount of sea level rise to its present level about 5000 years ago, as the ice in the polar regions melted in the present interglacial phase. The present rocky surface of the atoll is currently emergent just above sea level in the four small islands. On this land surface, one can see rocks that show alternations from limestone, formed under marine conditions when sea level was high, and fossil soils and erosion surfaces that formed when sea level was low. When global climate was warm, the land area of the atoll was periodically covered by the rising sea, and on at least two (possibly three) occasions the atoll was completely submerged, drowning any land animals then present. The animals that live on the islands today must have arrived by sea as colonizers, possibly by clinging to floating logs—a perilous journey, for sure, but one that must, from time to time, have successfully transplanted animals from the mainland or from neighbouring islands.

Despite its remoteness from land, and until recently its isolation from humans too, Aldabra is not a barren landscape. It is populated with a range of creatures, including most notably about 100 000 giant tortoises, *Aldabrachelys gigantea* (Figure 37). In captivity the tortoises can weigh in at 250 kilograms, though in the wild they are normally about 150 kilograms.

Figure 37. A female Aldabra tortoise (*Aldabrachelys gigantea*). This particular animal lives on Curieuse Island in the Seychelles.

Fossil bones of the giant tortoises of Aldabra show that their ancestors colonized the atoll perhaps 100 000 years ago. Yet older bones of giant tortoises are preserved in the fossil soils sitting atop the earlier limestone layers, together with the bones of crocodiles, lizards, and birds. Aldabra, then, has been reinvaded by the tortoises, and other animals, several times during the Pleistocene as sea level changed, in between the episodes of island drowning. The fossil record of the tortoises is, in its own way, an archive of climate and sea level change as effective as—and arguably even more evocative than—that provided by the tiny foraminifera in the deep sea oozes. Each time the Earth's spin through space changed, ice melted in the polar regions, sea level rose, and the tortoises of Aldabra were cast adrift in the rising eastern Indian Ocean. However, they lived on elsewhere, and so could recolonize the island thousands of years later, as the Earth drifted further away from the sun's heat again, and ice formed at the poles.

The Aldabra tortoise population is currently recovering after the 19th-century inhabitants of the islands nearly hunted them to extinction. But the populations of giant tortoises on adjacent islands through this region are now mostly extinct. When the sea once again submerges Aldabra—this time through our burning of fossil fuels, and not by the action of astronomical cycles—there may be few sources of tortoises to repopulate the Aldabra of the far future, when sea level falls again. New populations of *Aldabrachelys* introduced to the neighbouring Mascarene Islands (which have high volcanic peaks, and so cannot be completely drowned by sea level rise), though, offer some hope for the giant tortoise.[76]

Aldabra is a good example of how Earth's past environments can be confidently reconstructed using the fossils of species that are alive today. These fossilized skeletons speak directly to us of such things as their ecology, the types of food they ate, and the temperature of the seawater they lived in. In going back more than 10 million years, though, few of the species living today are the same as those in the fossil record. Even if some modern species were present then, they may not have had the same environmental tolerances as their living relatives possess today. In yet older skeletal archives of Earth's past environments, living species are absent from the record.

It is still possible, though, to discern relatives of living species in the older geological record. Coelacanth fossils are found in Cretaceous (and yet older) strata, while ginkgo trees thrived in the Jurassic. Further back, in the Ordovician seas of more than 440 million years ago, there are distant relatives of the living *Nautilus*, and yet further back, worms are amongst the common animals of the Chengjiang biota of the Cambrian Period that are related—though not identical—to animals still found today. Before that, and not long after the beginning of life in Earth's oceans, there are the stromatolites, biogenic structures made by microbes that still form in a few places today. The similarities between organisms from many millions of years ago and living species suggest that there is a record of environmental information stretching far back into the deep time of Earth. Can these ancient skeletons be used as an archive of past Earth environment? One answer lies in the delicate growth lines of Devonian corals.

Keeping Time in the Devonian

The passage of Earth through the heavens has a strong influence on climate, and this is recorded in the skeletons of both foraminifera and tortoises. On a shorter timescale, a sense of the Earth's motion may be obtained simply by watching the sky at night, as the stars and planets drift across the heavens. This motion, the Earth's spin momentum, began long ago, at the time that our solar system formed from a spiralling disc of condensing gas and dust. But the energy of that spin is gradually dissipating through the tidal effects generated by the Earth–Moon system, with part of that continuous, ongoing energy expenditure being felt as the ebb and flow of the tides. As time passes, the average day becomes slightly longer, so that over each passing century, the day becomes about 2 milliseconds longer. Eventually, the Earth's spin will slow, and the planet will begin to wobble—quite dramatically—on its axis, like a spinning top losing its energy. As this rate of dissipation of energy can be calculated, it is possible to estimate the length of a day and how many days there used to be in a year, going backwards in time. The further back, the faster the Earth should have been spinning relative to today, and the shorter each day would have been.

The skeletons of many marine creatures are known to record seasonal changes in weather and tides, as evident from the ocean quahogs of the North Atlantic. But if the Earth was spinning faster in its younger life, then there should have been more days in a month, and more months in a year. Could these changes be captured in the growth patterns of very ancient skeletons? Two geologists working in the 1960s, Colin Scrutton, then at Oxford University, and John West Wells at Cornell University in

New York State, set out to provide a test of the changing motion of the Earth using fossils. Wells was accustomed to studying the growth patterns of modern corals, having worked in the Pacific Ocean during the 1940s, including on Bikini Atoll, the site used for testing nuclear bombs by the US military. The two men examined well-preserved fossil corals from the Devonian Period, a time stretching back from 359 to 419 million years ago.

Corals grow their skeletons from calcium carbonate, materials that are readily available in seawater, adding new layers that appear as a series of growth lines in the wall of the coral skeleton. Wells and Scrutton totted up the number of growth lines on the skeletons of an extinct group of corals called the rugose corals (Figure 38). These corals mostly lived a solitary existence on the seabed, never forming the giant reef systems of modern corals. But many individual rugose corals nevertheless grew large, to look like inverted horns of a Viking's helmet, and were clearly long-lived organisms. Though rugose corals were common in the seas of

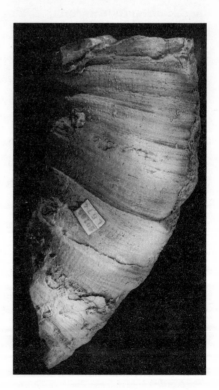

Figure 38. The rugose coral *Caninia*, from the Carboniferous of County Sligo, Ireland. From tip to tip the specimen is 10 cm long.

the Palaeozoic Era, they had become extinct at the end of the Permian Period, about 252 million years ago and so the question remained, could a skeleton belonging to an extinct group of corals from more than 360 million years ago really record the length of a Devonian year? Wells' careful examination of the growth patterns in the coral skeletons was published in the scientific journal *Nature* in 1963,[77] and it suggested a Devonian year of about 400 days. A year later,[78] after corresponding with Wells, Scrutton confirmed these observations showed, based on fossils from both North America and Europe, that rugose corals from the Devonian Period added increments to their skeletons in groups that reflected lunar months, with an average of 30.6 days, and therefore with 13 lunar cycles in a year.

Could these techniques for discerning ancient months and years be taken back yet further in time, into the Precambrian when the range of skeletons available for analysis dwindles? Some scientists suggested that the calcium carbonate layers of stromatolites might reflect daily growth layers, or responses to tides, or both. Suggestions of a Precambrian year with about 425 to 450 days have been made, but there is no consensus on whether stromatolites really do record these trends.

The jury is out on the viability of stromatolites as timekeepers, but this does nothing to devalue the work of Wells and Scrutton. Their papers spurred the geological community to examine a whole range of different fossils for their potential environmental signature. Most importantly, the two men's coral work showed that even when organisms were long extinct, the fossil evidence from skeletons retained a powerful narrative of astronomical influence. The field was now open to interrogate the climate signature of some of Earth's most ancient animal skeletons.

The Warm Seas of the Ordovician

It soon proved obvious that fossil corals were not the only useful archive of environmental change in the Palaeozoic seas, and that skeletons made from a diversity of different materials, including carbonate, silica, apatite, and complex organic materials, might yield important stories too. Mostly, analysis showed that the original chemistry of these ancient fossils had been changed by heat and pressure, as they lay deeply buried in the Earth's crust, the original minerals being recrystallized and the original chemistry being scrambled. However, amongst these ancient fossils, the teeth-like conodonts were seen as more resilient to such change, and scientists sought to use them to measure the temperature of the seas in which these extinct early vertebrates lived.

Modern tropical oceans typically have sea temperatures of around 30°C, though in a few more land-locked seas, temperatures may rise a little higher. Warmed by the sun's rays, the heat of the tropical seas is carried to higher latitudes by ocean currents, and by the air circulating above. This redistribution of heat in the open oceans helps prevent the tropics from getting too hot, and the polar regions from getting too cold. This engine of ocean heat transfer seems to have functioned even during the global greenhouse conditions of Cretaceous and Paleogene times, with tropical sea temperatures only sporadically rising much above 30°C.

Hays, Imbrie, and Shackleton (and their many colleagues) had plotted a clear path in establishing the climate history of the time since the dinosaurs. However, that path became more overgrown as scientists looked deeper into the geological record—and then, seemingly, vanished altogether. For all of the marvellous fossils recovered from the early Palaeozoic strata of the Burgess Shale and Chengjiang Biota, the picture of early life that these provided was accompanied by scant information on the degree of warmth of the seas in which these animals lived. Then, about a decade ago, a team of scientists led by Julie Trotter, a geologist at the Australian National University, set out to clear this path. They examined the oxygen isotope signatures from Ordovician conodonts living in the tropical regions between 443 and 489 million years ago.[79] To make their calculations they had to make some bold assumptions about the chemistry of Ordovician seawater. And they also had to be sure that the conodonts had not been recrystallized in the more than 440 million years that had elapsed since they formed. Trotter and her team worked to overcome these problems, and found a dramatic trend, with the chemistry of the oldest conodonts suggesting surface ocean temperatures of about 40°C. These very warm tropical seas seem to have persisted for several million years, and they may have characterized much of the Cambrian Period before. Only some 25 million years after the beginning of the Ordovician Period did the conodonts begin to show tropical sea temperatures more like typical modern values of around 30°C.

The cooling trend that Trotter and her colleagues had identified coincides with a great flowering of marine life that is called the 'Great Ordovician Biodiversification Event', or, as it has come to be known now by palaeontologists, 'GOBE'. This event did not introduce new types of animals, but rather was a diversification of species of the major animal groups that had originated in the late Precambrian and Cambrian seas. GOBE is also associated with an increase in creatures that used calcium carbonate to make their skeletons. This trend can be clearly seen by, say, examining a kilogram of sedimentary rock from the early Ordovician and noting that it might contain the skeletons of a few ostracods belonging to a couple of species. In contrast, a grab sample of fossilized seabed from some 30 million years later will typically contain a thousand or so ostracod skeletons, belonging to 30 or more species. GOBE also involved a race to colonize the plankton, with skeletonized groups such as the nautiloid cephalopods, arthropods, and graptolites all becoming well established and diverse. The changes to the ocean ecosystem that ensued increased the food supply for animals living on the seabed, as larger zooplanktonic animals, such as arthropods, delivered a continual stream of organic-rich poo to the seabed. This rain of nutrients, which has been termed the 'biological pump' (oceanographers also refer to it as the 'faecal express'), may have had a fundamental effect on the ocean's carbon cycle too: one that may have helped to cool the Ordovician seas. The pump forms a major part of the Earth's life support system, exporting carbon from the surface to be stored in the deep ocean, and so it therefore prevents carbon, as carbon dioxide, from building up to excessive levels in the atmosphere.

The increasing range of planktonic creatures that occupied the Ordovician seas might therefore be linked with the cooling ocean temperatures. These creatures would have helped the biological pump to function by trapping carbon in their bodies and then delivering this to the seabed, in turn reducing the level of carbon dioxide in the atmosphere. The negative greenhouse effect so produced would have cooled the oceans too—and so put in place the conditions for the huge diversification of life in the sea that was GOBE.

Sea temperatures more or less then stabilized for some 20 million years, but at the end of that period the Earth was plunged into a severe icehouse climate state. As ice grew enormously on South America and southern Africa, which were then on the South Pole, sea level dropped precipitously, exposing the continental shelves and triggering the first of the big five mass extinctions of the past half-billion years. The cause of this event is still mysterious, but current suspicions do not place animal bodies, and animal skeletons, as a principal cause of cooling. Rather, large-scale collision of continents, the uplift of giant mountain chains, and changing patterns of ocean circulation are considered possible causes.

Such climate riddles are bound up with ancient geography—and the changing patterns of the geography of our planet were, famously, to provide a cornerstone of plate tectonic theory. As oceans open and close, and as continents are separated and coalesce once more, the animals and plants caught up in these planetary changes respond by changing their evolutionary pathways. Their skeletons are an archive of the movement of the continental and oceanic plates too.

Skeleton Coasts

In the 16th century, the great Flemish cartographer Abraham Ortelius noted the good fit of the continents of South America and Africa (Figure 39).

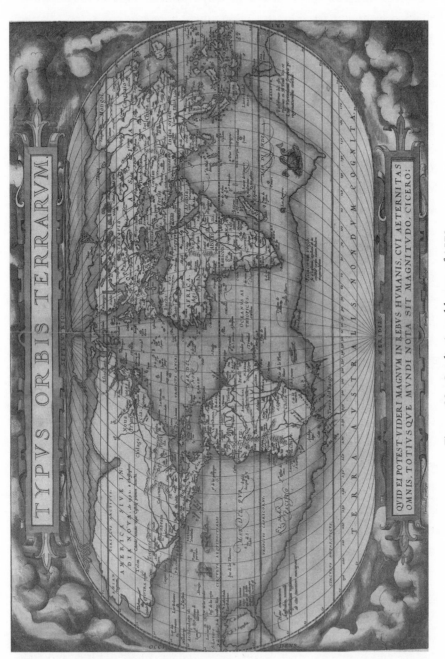

Figure 39. Ortelius' world map of 1570.

He was able to do this because his maps were the best of their kind in the world, being founded on more than a century of trans-Atlantic shipping. Ortelius thought that the two sides of the Atlantic had been ripped asunder by some catastrophic event, but he did not have the geological evidence to support this notion, not least because the science of geology had not yet been born. The idea that the continents might somehow move apart at the surface of the planet found support slowly, as geological data grew gradually. And only in the early part of the 20th century did it resurface again as a serious idea, through the work of German scientist Alfred Wegener. It was then in the mid-20th century, when technology at last caught up with Ortelius' insights, that physical evidence was found that oceans could widen.

The 1960s were a heady time for geologists. The decade saw the publication of Marie Tharp and Bruce Heezen's map that clearly showed for the first time the structure of the deep ocean seafloor, revealing a gigantic line of submerged mountains, of the mid-ocean ridge, extending down the middle of the Atlantic. In tandem, Canadian John Tuzo Wilson and colleagues recognized that ocean crust showed symmetrical magnetic stripes, in parallel on either side of these mid-ocean mountains. These stripes, like a gigantic barcode, had been detected from ship-borne magnetic surveys, and they showed conclusively that new ocean crust, formed at the mid-ocean ridges, as it cooled, fossilized a remnant of the orientation of the Earth's magnetic field within it. As newly formed crust moved away from the ridge, the crust forming behind it recorded the periodic switches in the polarity of the Earth's magnetic field, as the north and south magnetic poles changed places.

It took until the mid-20th century to understand how the topographic and magnetic properties of the deep ocean reflected the movement of oceans and continents. However, the fossil archive that was to provide its own stories of evolving oceans and continents was already well established, and had been quietly and systematically accumulating since the mid-19th century.

Amid the majestic mountains and glens of Scotland, fossils from hundreds of millions of years ago showed that the rocks were completely different from those of similar age in England and Wales. Therefore, long ago, these two regions of the today's British Isles must have been separated by a formidable geographical barrier. Little sense of this geographical dislocation is apparent on the short drive from the English Lake District in Cumbria to the hills of the Southern Uplands of Scotland. Nevertheless, the fossils clearly show that 500 million years ago, these landscapes were an ocean apart.

The great ocean that separated these land masses existed over more than 100 million years, from its birth in the late Precambrian, to its demise in the early Silurian Period. Its story is recorded in the skeletons of animals in rocks from either side of this great seaway, in North America and continental Europe, and its demise, when those land masses finally collided, can be seen in the mountain chains of the eastern United States and Canada, the highlands of Scotland, and those of Norway too.

Two pioneering 19th-century geologists of the British Geological Survey, Benjamin Peach and John Horne, first developed these ideas. 'Peach and Horne' are legendary in the history of British geology, particularly for their masterly work in unravelling the structure and history of the ancient, tectonically deformed rocks of the remote northwest Scottish mountains. The two geologists complemented each other near perfectly. Ben Peach, a genial man, was inspirational in the field, having an extraordinary ability to intuit the complex three-dimensional structures of these rocks, and draw them out as sketches and geological cross-sections—but he was often disorganized, slow with his administrative duties, and was reluctant to put pen to paper. John Horne was logical, organized—and a good and prolific writer. Between them, they produced an account of these rocks that is still the starting point for all subsequent studies.

Peach and Horne noticed how the Scottish bedrock was more akin to that of Greenland and North America, now thousands of kilometres away across the Atlantic Ocean, than it was to that of its English neighbour. This far northern landscape is founded on Lewisian Gneiss—ancient, highly deformed rocks named after the Hebridean island of Lewis that are now known to be nearly 3 billion years old. There is nothing like these rocks in England—but they are akin to the ancient rocks that form the bedrock of Greenland and North America. Upon these ancient rocks, and separated by an irregular and weathered surface called an unconformity, which represents a huge break in the geological record, lie the younger Precambrian rocks of the Torridonian, which make dramatic highland landscapes. But it is the rocks that sit above the Torridonian, separated by another unconformity, that yield telltale skeleton evidence of Scotland's North American heritage.

Peach and Horne, in making these deductions, were generous in acknowledging Survey palaeontologist John William Salter, who half a century earlier in 1859 had first recognized how fossils also pointed to this relationship. Salter was one of the great palaeontologists of the 19th century. He had worked alongside those two influential British geologists who between them carved out the early Palaeozoic timescale, Englishman Adam Sedgwick and Scotsman Sir Roderick Impey Murchison, who founded the Cambrian and Silurian systems, and then warred over where the boundary should lie between them until that other great geologist (and practical diplomat) Charles Lapworth set up the Ordovician System to separate both the Cambrian and Silurian and their (by now deceased) protagonists.

Salter, in the midst of these discoveries and battles, developed a profound knowledge of a wide range of fossils, including trilobites, making a name for himself as someone who could identify these fossils from as far afield as the Himalayas and Australia. Cambrian cephalopods and gastropods of northern Scotland, he stated, had more affinity with North American forms than with European ones. He then had no inkling of the workings of plate tectonics, but, like Abraham Ortelius, he is a key part of the story that was to play out a century later. Salter, unwisely, resigned his post at the Survey one year short of the term needed to secure a proper pension. Plagued by increasingly poor health and by financial woes, he jumped to his death in the Thames in 1869, leaving a widow and seven children. A year later, no less than Thomas Huxley wrote his epitaph in the presidential address he gave to the Geological Society, noting Salter's boyish enthusiasm for all things palaeontological.

Salter's recognition that the fossils of the Northwest Highlands had a North American affinity was just the tip of the iceberg of a torrent of palaeontological data that showed this was so. From the Northwest Highlands to the Midland Valley of Scotland, the Cambrian and Ordovician fossils all showed the same signature. In Girvan, for example, a small fishing town on the western coast of the Midland Valley, the Ordovician mudrocks and limestone have fossil species—including ostracods—that are well known from Virginia in the United States.

Across the Solway Firth to the south, the Cambrian and Ordovician fossils are distinctly different, at least until the end of the Ordovician. Thereafter, the fossils either side of the Solway gradually converge in their biological affinities, as a narrowing ancient ocean, the Iapetus Ocean, was destroyed, subducted beneath the continents on its northern and southern margins. The first contacts are seen in the fossils of animals that lived in offshore deep marine settings of Iapetus that had the capacity to cross the narrowing sea. Later, as the ocean disappeared, and only narrow seaways remained, the animals from very inshore settings also began to increasingly resemble each other.

As the Iapetus Ocean closed, thrusting up its great mountain chains between the ancient terranes of Scotland, Wales, and England, the land was changing colour too. For countless millennia it had been weathered rock, perhaps with a thin veneer of bacteria and algae. Now, gaining a toehold slowly at first, in low-lying damp coastal plains, plants were invading. A once silicate-grey landscape was turning green. And the green shoots of these plants would, in turn, provide an eloquent archive of information about terrestrial environmental change over geological timescales.

The Long Sleep of the Antarctic

Skaar Ridge lies 84° south, on the flanks of Mount Augusta in the Queen Alexandra Range of the Trans-Antarctic Mountains. It is a small ridge of rock, marked by a cairn of stones, which extends for just a few miles towards the mighty Beardmore Glacier. The icy landscape here is blanketed in darkness for 6 months of the year, and even during the height of summer, temperatures stay well below zero. Long ago, in the warm climate of the Permian Period, plants grew here well beyond the southern latitude of modern tree growth. And these trees were of a type that was widespread across many parts of the world, from Africa to India and Australia.

The Skaar Ridge site has yielded many fossil plant specimens of Late Permian age.[80] These fossils are part of a long tradition of geological collecting from this part of Antarctica. On 12 February 1912, on route back from their failed attempt to be the first human visitors to the South Pole, Robert Falcon Scott's ill-fated expedition stopped off at the top of the Beardmore Glacier and, observing some interesting rocks, decided it would be a good place to look at the geology. This rock-hunting sojourn might seem a bizarre act of self-confidence, given that within a few weeks Scott was to write the final entry in his diary. But a day's collecting added some '35 lb' (about 16 kilograms) of rocks to the expedition's load, and given how little was known about the Antarctic continent at that time, the samples were akin to returning rock from the surface of the Moon. Scott's expedition abandoned some of their field gear in an attempt to lighten their load on the return journey, but they clung on to these scientifically important specimens to the end, a measure of true greatness in the face of severe, indeed fatal, adversity. Amongst the rocks that Scott's expedition collected was a fossil of the Permian plant *Glossopteris*, which now resides in the Natural History Museum in London. *Glossopteris* is proof that once upon a time Antarctica had been linked with its African, Indian, and Australasian cousins in one mighty landmass.

The rocks on Skaar Ridge preserve a fossilized peat, and within this are the leaves and stumps of glossopterids, the latter so well preserved that individual tree rings can be discerned after more than 250 million years. These fossils tell a story of trees growing at high latitude in a warm world quite different from that of today. Each year that a tree grows, it adds a new circle of growth to the girth of its trunk. This growth occurs in a layer of cells just beneath the bark of the tree, and in modern temperate zone trees, where there is a strong contrast in seasonal growth, these rings are clearly defined. A dendrochronologist—one who studies the signature of these rings—can distinguish wood formed in the spring (early wood) from summer growth (late wood). The width of the ring may vary between a good year and a bad, drought-prone year. Tree ring changes accumulating over time can be compared across trees, and rather like the techniques of measuring growth increments in bivalves such as Ming, a long record of seasonal variation can be built up from the overlapping records of progressively more ancient trees, and this has been reconstructed back in time over many thousands of years. The idea is not a new one. The French polymath and genius the Comte de Buffon had already in the 1730s, with his colleague Henri-Louis Duhamel du Monceau, observed the impact of the severe winter of 1709 on tree ring growth.[81]

What then of wood found in the very ancient fossil record? Might this reveal something of the environment the trees were growing in so long ago? At Skaar Ridge the fossilized leaves of Glossopteris occur in dense mats showing that, just as deciduous trees do today, these high latitude trees dropped their leaves as winter approached. But the glossopterid trunks show something very different, with almost no late wood, and instead tree rings dominated by early wood. This is a pattern not seen in living trees, and it suggests a transition into winter dormancy that occurred very sharply. The glossopterid trees of Skaar Ridge show that as the sun dipped below the horizon at the start of the long Antarctic night, the trees fell quickly into a deep sleep, only to reawaken when, after 6 months of darkness, the sun once again appeared above the horizon.

A Brief Royal Epilogue

The excavators dug through the car park surface opposite Leicester Cathedral at the point where, spookily, a crudely painted 'R' on the tarmac took its place in a series of alphabetically marked parking spaces. Beneath the ground, in a crude and hastily made grave, there was a male human skeleton (Figure 40). The skull showed the marks of the sword blows that had killed him, the back of the cranium hacked away. These bones held many secrets, and one was to identify him. DNA extracted from tiny samples of them showed that these were the long-lost mortal remains of the last Plantagenet King of England, Richard III. The story of his discovery and forensic analysis, by a team of archaeologists and historians from the University of Leicester, the city council, and the Richard III Society, is now a classic. And the bones told their stories in more ways than one.

Figure 40. King Richard III of England, shortly after the discovery of his remains beneath a car park in Leicester.

Richard III has had a very bad press. As vividly portrayed by William Shakespeare, he personified villainy and ambition, his outward appearance of a hunchback with a withered arm matching his moral darkness as the murderer of the young Princes in the Tower, justly meeting his comeuppance on Bosworth Field against the heroic Henry VII, the first of the Tudors. Was it really so? Truth, mused Josephine Tey, is the daughter of time,[82] a thought that allowed her to pursue the notion that perhaps it was not Richard, but the first Henry Tudor, who murdered the young princes, and then successfully covered his tracks, with Shakespeare, more than a century later as the Tudor regime still held sway, caught up in creating some of the most atmospheric and convincing propaganda ever written.

What do the bones say? The physical deformity clearly was not a rumour. Richard III suffered from severe scoliosis of the spine, something that may have given him a hunched posture, perhaps with one arm held higher than the other. The bones betray other parts of his history. The nature of the evidence would have mystified Shakespeare—the pattern of isotopes of lead, strontium, nitrogen, oxygen, and carbon—but the stories they told would surely have allowed him to embellish his play with vivid, and true, detail to illuminate the king's life. In some respects, time's daughter can faithfully recall fragments of the past.[83]

The tooth enamel of Richard III formed early, just as it does in all normal mortals. As it formed, it absorbed elements of the landscape amid which the boy grew up. The strontium isotopes reflect the local food, which in turn reflect the local geology upon which wheat grew and sheep and cattle grazed, and the oxygen isotopes reflect rainfall patterns and their interaction with geography. The patterns of these isotopes, extracted from the king's teeth, clearly show that Richard moved away from Northamptonshire, where he had been born, at about the age of seven, to a place of different geology and heavier rain, probably somewhere in western England (perhaps Ludlow on the Welsh Borders, the authors of the study suggested).

A little later in life, patterns of these isotopes in bones that carry on growing past childhood—dentine in the teeth, the femur—show that the king-to-be moved back to the drier areas and the geology of the east. The lead in his teeth showed that, when he lived, the air, soil, and water were no longer as pristine as they had been in the time of his ancient ancestors: the traces of pollution from metal smelting in the region are evident.

The king's rib showed another pattern. This bone renews itself quickly through life, and provides clues to his last few years, already as king. The carbon and nitrogen isotopes here betray a rich diet, with wildfowl such as swan, egret, and heron and freshwater fish figuring prominently—and wine, too: the analysis of his bones was the first example where the pattern of nitrogen isotopes has been linked with copious wine drinking. The feasting and banqueting of a royal life in medieval times were no myth, it seems.

As to the moral character—or guilt and innocence—of Richard III, the isotopes remain silent. The daughter of time, here, remains enigmatic. But one can at least raise a glass to the long-dead monarch, knowing that to be an entirely fitting farewell.

9

FUTURE SKELETONS

The asteroid that slammed into what is now the Yucatan Peninsula of Mexico, 66 million years ago, was about 10 kilometres across, or the diameter of a moderate-sized city. The energy instantly released was about a billion times that of the Hiroshima and Nagasaki atom bombs combined, creating a crater some 180 kilometres in diameter. The powerful shock waves, rippling around the globe, may even have helped provoke bursts of volcanism in India, then exactly on the opposite side of the globe. Enormous tsunami raced across the oceans, firestorms were triggered, enormous amounts of soot and dust in the atmosphere probably triggered some kind of 'nuclear winter' cooling for years, and the oceans were acidified from the sulphates released by the impact.

The exact kill mechanisms are still hotly debated by scientists, but kill they did, and kill most effectively. When the dust settled, there were scarcely any animals alive on land larger than 25 kilograms. The non-avian dinosaurs[84] had been wiped out, and the ammonites and belemnites had disappeared from the seas. It was the fifth major catastrophe to strike life on Earth in the past half-billion years. It was not the most severe of these crises—the volcanism-triggered Permian–Triassic extinction event, 252 million years ago, takes the palm for that—but it was almost certainly the most abrupt. Life had to reorganize itself, once more—and design new skeletons to replace the spectacular ones that had been so suddenly discontinued. There are many stories surrounding this particular reorganization, but the story that is in our bones, so to speak, is how

mammals emerged from the shadows of the dinosaurs, after that catastrophe, to dominate both land and sea.

The mammals' sojourn in the shadows had certainly been an extraordinarily long one. Throughout the whole of the Jurassic and Cretaceous
periods (and in part of the Triassic Period before that), they were present
on Earth. This is clear, even though most of the features that we associate
with mammals—the fur, sweat glands, mammary glands, the way they
care for their young, and so on—do not fossilize easily. However, the
early mammals can be recognized from their specialized teeth, which
form a familiar package of incisors, canines, premolars, and molars.
Such teeth—the hardest part of their skeletons—consistently turn up in
strata in which there are dinosaur bones. Their equally consistently
diminutive size, though, indicates that mammals then were small,
mostly varying from the size of a mouse to that of a domestic cat.
The largest known seem to have been about as big as a badger, like
Vintana, which lived in what is now Madagascar in Cretaceous times,
and the similarly sized *Repenomamus*, also Cretaceous, from China. And
of those that survived the end-Cretaceous meteorite impact, none
seems to have been larger than a rat.

That newly emptied ecospace, though, offered possibilities to those
diminutive survivors, now that the dinosaurs were gone from the top
spots of the food chain. From those small beginnings came a cornucopia
of new mammals that evolved into an extraordinary range of forms. Even
the first 10 million years after the disaster, the time of the Paleocene
Epoch, showed a transformation. From the rat-sized mammal survivors,
larger and more diverse forms quickly began to evolve. Less than a
million years after the impact, at least one mammal, *Wortmania*, had
reached 20 kilograms in bulk. Then, there came heavy-boned large
herbivores such as *Barylambda*, in shape something like an early version
of a giant sloth that could weigh more than half a ton. There was the
fearsome-looking *Titanoides*—the size of a pig, with huge sabre-like
canine teeth and clawed feet, it was nevertheless likely to be a herbivore,
digging for roots. Carnivorous mammals had appeared, strange animals

like *Psittacotherium*, the 'parrot beast' with massive front teeth drawn out in beak-like fashion.

That was just the first 10 million years. In the succeeding Eocene and later epochs, the array of descendants from those original rat-like ancestors continued to diversify. There was the 2-ton Eocene *Uintatherium*, about the size of a rhino or hippo, with a little of the appearance of both. It was the kind of beast that may have been preyed upon by the mysterious *Andrewsarchus*, known only from a single metre-long enormous lower jaw, embellished with fearsome teeth, discovered in the wilds of Mongolia nearly a century ago—perhaps the largest carnivorous land mammal ever. These were impressive beasts, but were later eclipsed on land by *Indricotherium*, whose 20-ton bulk brought it near to the size of the giant sauropod dinosaurs, and in the oceans by the whales, including the blue whale, the largest animal ever to have lived on Earth. These, and the horses, camels, elephants, tigers, gorillas, tapirs, and many more—all (including the blue whale) were derived from those tiny furry rat-like ancestors in just 66 million years. Their bones, when unearthed now and again by the ancient Greeks and Romans, were often interpreted as those of giant humans—either ogres or heroes. A source of awe and fascination, and revered as sacred relics or as talismans to bring good fortune in battle, they were among the celebrity items of those times, and a good discovery could spark off a bone rush.[85]

The evolved animals hang, of course, from an evolved bony framework. Consider the jaw of the rat-like ancestor, a couple of centimetres long, weighing a few grams, and ending in sharp teeth. Then fast forward a few tens of millions of years, and then take the 107 bus route in Barnet, London, stopping off at Wood Street by a house called The Whalebones. There is an entrance arch to the driveway—which is made of a pair of real blue whale jaws anchored into the ground—their original owner having been hunted to death in the South Seas. Approaching 7 metres long and weighing about a ton and a half, in life they would not have had teeth, but sheets of baleen for the filtering of plankton from seawater.

The strange and rather grisly memorial in Barnet gives pause for a good deal of thought, but one conclusion might be that it shows what evolution can do to a skeleton, over a time span of about 1% of the age of the Earth. Just to hammer home the point of the sheer variety, and quirkiness, of the process, one might look at a smaller, though no less remarkable, cousin of the blue whale, the narwhal. This is one of the toothed, rather than baleen whales, though it is in the most literal sense a singular phenomenon. Virtually all the teeth have been lost in its evolution, save one of the canine teeth. In the male (and in some females) this tooth grows straight out in front of the animal, reaching a length that can exceed 3 metres, to become the defining characteristic of this 'unicorn of the sea'. The giant single tooth is a multipurpose tusk, crammed with nerve endings that transmit information about water conditions to the whale's brain, and, it has been suggested, to other whale brains, too, when whales rub tusks together. The tusk, too, can be used to strike and stun small fish that are the whales' prey. This is a science-fiction skeleton—but one that exists in reality, produced by nothing more than Darwin's marvellous mechanism and modest amounts of geological time.

Preparation for a Sequel

It is hard to imagine, now, what the Earth's biological richness used to be like, until geologically recently. We have an abundance of vertebrates on land right now—perhaps an abundance that has never been matched at any time in our planet's history. But this abundance is now concentrated, just as uniquely, within a tiny, all too familiar, set of creatures—us and our farmyard animals. The wild vertebrate land animals of the world are now pushed to the margins, their combined mass amounting—as the scientist Vaclav Smil has estimated[86]—to less than 5%—perhaps even under 3%—of the mass of us and our selected prey animals.

Most of the species that were present a century or two ago are still here. But that number, as regards large mammals, was already reduced.

Between 50 000 and 10 000 years ago, the number of large animals was hugely depleted, with the loss of some 90 genera of animals weighing more than 44 kilograms,[87] approaching half of the total. Among the disappearances were sabre-tooth tiger, giant sloth, and mastodon in the Americas; mammoth, woolly rhinoceros, auroch, and giant deer vanished from Eurasia; and the rhino-like *Diprotodon* and giant wombats disappeared from Australia. Climate change may have been implicated, but climate had changed many times over the previous few million years, without sparking extinctions. Almost certainly, this wave of extinctions was humanity's first significant mark on the Earth, via a formidable, honed skill at hunting.

In the 11 700 years of the Holocene, the scale of these extinctions was only matched when humans discovered new terrain: New Zealand, for instance, had been inhabited by several species of that large flightless bird, the moa, the largest well over 3 metres in height, preyed upon only by the huge Haast's eagle. At about AD 1300, the Maori people arrived, and little more than a century later all moas—and the eagle—had disappeared.

The remaining large wild animals have been confined to what is left of the Earth's wilderness. Now, that wilderness is receding quickly, with humanity's extraordinary growth, from about 1 billion in AD 1800 to about 3 billion in 1950, to 7.35 billion now, and heading towards 11 billion by mid-century. Extinction rates globally have gone up by an order of magnitude over the last century, and may now be something like a thousand times over background levels. If these trends continue, how long to the next full-scale mass extinction? Anthony Barnosky and his colleagues' work suggests that, with business as usual (and not including the effects of climate change), it will take just a few centuries for a mass extinction, rivalling that at the end of the Cretaceous, to take place.[88]

Let us hope that this tragedy (for there is no other way of putting it) will not come to pass. But, currently, the growing pressures show no signs of abating. It seems more likely than not that, geologically very soon, there will be another emptying of ecospace. And humans themselves—living more quickly and more dangerously than any other species, might be among the casualties. Taking the long view ahead—what then?

Bones of the Sequel

The palaeontologist Dougal Dixon has been called the founder of speculative evolution. He imagined a world shorn of many species because of man, and then shorn of man—and left for 50 million years. He populated this world with a plausible but entirely imaginary new suite of animals, evolved from the survivors that were left *After Man*,[89] as the first of his books on this theme was titled. If the elephants became extinct during the brief human empire, he said, perhaps their ecological position could be filled via the evolution of some of the few surviving antelope. The result he called the gigantelope, an elephant-sized animal where the antelope horns have stretched out and coiled forward to take the place and the function of the tusks. With true anteaters extinct, their niche he imagined to be taken over by the turmi, survivors of pigs, their snout bones extended and with horns outwardly turned to dig into termite nests, and a lower jaw that has lost teeth, becoming a narrow channel for the long, termite-gathering tongue. Among Dixon's predators is the sinister-looking strider, a descendent of a cat adapted with limbs greatly extended and armed with grasping claws to swing through the forest branches in search of prey.

He provides many more examples. It's a beautifully imagined (and drawn) picture of what the biology of the far future might be, over a time roughly similar to that of the Cenozoic Era. Indeed, given what we know of the extraordinary animals of that era (still, remember, the one in which we live) it is a little conservative—Dixon does not posit anything quite as extreme as the something like 100 000-fold increase in mass of the jaw (and total restructuring in shape) from that of the end-Cretaceous mammal survivors to that of the blue whale. Truth can be stranger than fiction, even properly scientifically imagined fiction.

Dougal Dixon focused on the mammals, to provide a 'futurized' mirror image of their spectacular Cenozoic evolution. To obtain a more rounded view of future skeletal evolution in a human-free world, we need to spread our net more widely among the other groups of organisms. Among the reptiles, the snakes are managing to stay among the survivors of the human onslaught, and some even appear to be the winners, like the pythons introduced into the Florida Everglades, now exploding in population and seemingly unstoppable as a voracious predator. Will such redistributed survivors ever evolve into a snake the size of the Paleocene *Titanoboa*? Reconstructed from a few enormous fossil vertebral bones, this snake is thought to have been in excess of 12 metres long and to have weighed significantly over a ton.

Of course, there are the invertebrates to think of, too.

The Next Reef Gap?

As the year 2015 turned to 2016, the Pacific Ocean began one of its periodic changes. Warm water began to spread from its western part to its eastern part. This is El Niño, part of a climate oscillation that has taken place every few years for thousands, and probably millions, of years. It brings with it flooding to the western Americas, and drought in the west Pacific, India, and Australia. It has brought collapse of anchovy fishing off Peru, and famine to Africa and to Europe too, may have brought about the demise of pre-Columbian cultures in South America, and perhaps helped spark off the French Revolution of 1789. But in early 2016, there was another victim. It was the Great Barrier Reef.

As the waters off eastern Australia warmed to levels probably not experienced there for many tens of thousands of years, the corals of the Great Barrier Reef began to suffer heat stress. This triggered their classic reaction: expulsion of their zooxanthellae, which removes both their colour—bleaching the formerly vividly coloured coral animals to a ghostly white—and a major source of nutrients. The more severely bleached corals died soon after.

Much of the damage was in the north, where over 60% of the coral cover was bleached. The south—normally more vulnerable, and more damaged by human activities in general—was largely protected by a vagary of the weather: the passage of a hurricane there, early in the year, cooled the waters sufficiently to prevent most of the bleaching. Without that freak event, the damage would have been much worse.

As we write, in 2017, the damage did just become worse. Although El Niño is receding, the waters in this Australian summer were still hot enough to severely bleach corals in the mid-section of the Great Barrier Reef. In these 2 years, some two-thirds of the Great Barrier Reef has suffered severe bleaching. It is not just this iconic Australian reef that has suffered. Again, as we write, the warming waters in the South China Seas are decimating the corals. Around Dongsha Atoll, there was a local heat amplification effect, raising the temperature of the water by as much as 6°C above normal, and some 40% of the reef corals have died. The whitened reef skeletons are now darkening as the coral tissues putrefy.

Some 50% of the world's coral reefs have already been lost by a combination of heat, pollution, and disturbance. The heat is now taking on the main kill role, far earlier than expected. A few years ago, scientists were more concerned about the effects of increasing ocean acidification, as human-produced carbon dioxide continues to dissolve into the seas, over the coming century. Now, the (real) threat of acidification has been eclipsed by the effect of rising temperatures, with projections that 90% of the reefs will be gone by mid-21st century—and the warming, of course, is not likely to stop after 2050.

Temperatures look set to rise still further over coming decades. Now, bleached coral, if it has not been killed off completely, can recover, given sufficient time—a decade or two—to grow back. But with carbon dioxide levels and temperature set to rise still further (there are no signs yet of a slowing in this trend, let alone a reverse of it), the prospects of more bleaching events look high, and the prospects for coral reefs as functional ecosystems—and as producers of skeletal rock in large amounts—look bleak. Only a quick and dramatic slowdown in carbon emissions is likely to prevent this, and again, as we write, this does not seem likely. Rapid collective action on global environmental problems is currently not a well-developed human trait. The planetary mega-skeletons that are coral reefs look likely to disappear once more—the last time was 55 million years ago—from the Earth. We hope we are wrong, but on current evidence, reef extinction seems much more likely than not. In the words of one scientist, the reefs might already be zombie ecosystems—the living dead of the oceans.

Perhaps one might seek scant consolation in looking very, very far ahead to see what might ultimately arise, phoenix like, from the ashes of the world's current reef systems. So—if today's reef-building corals have been too badly damaged—or soon will be—to make a comeback, once Earth's temperatures begin to decline from the climate hyperthermal and ocean acidification event that is now beginning: what might take their place?

When the reefs died out 55 million years ago, they were replaced in many places by ecosystems dominated by those skeleton-making protists, the foraminifera, before the scleractinian corals eventually made a comeback—though these ecosystems had nothing like the strength and geometrical (and ecological) complexity that interlaced coral skeletons can provide.

We live, now, in an era of shelled molluscs, notably the bivalves and gastropods—a dominance that started when the dinosaurs walked Earth and that has only grown since then. Perhaps, therefore, some skeletons similar to those of the tube-like rudist bivalves of Cretaceous times will evolve again, and begin to build enormous collective structures on future warm and shallow sea floors. Perhaps the coralline algae—less sensitive to temperature than the delicate corals—can still form the leading edge of reef-like structures. Let us hope that we would not go back to a world of stromatolite reefs like that of the Precambrian—for that would mean that humanity would really have collapsed life down to its primeval basis.

Designer Skeletons

When Mary Wollstonecraft Godwin was just 18 years old, she had been the lover of the poet Percy Bysshe Shelley for 2 years. She, Shelley, and her stepsister Claire Clairmont went to stay for the summer of 1816 at a villa by Lake Geneva, to be with Lord Byron and his physician, John Polidori. The atmosphere was an intense mix of passions, intrigues, and intellectual adventure, focused all the more because this was the 'year without a summer', the eruption of the volcano Tambora the year before having affected the world's climate. Cooped up in the villa in the miserable weather, they amused themselves by recounting stories of the spirit world, and Byron challenged everyone to write their own ghost story. It was one of the most productive calls in the history of literature. Two monsters emerged that would shiver spines around the world from then onwards. Polidori wrote *The Vampyr* based on a Byron poem, and started the vampire horror story genre. And Mary Shelley (as she had started calling herself then), after days vainly seeking inspiration, had—between 2 AM and 3 AM on 16 June 1816—a vivid 'waking dream', and began to write *Frankenstein: Or, the Modern Prometheus*.

Frankenstein's monster is a classic tale of disastrous human hubris, as the brilliant but flawed Dr Frankenstein cobbles together the unfortunate creature from various body parts taken out of slaughterhouses and dissection rooms. Eight feet tall because the ambitious doctor struggled with the fiddly detail of anatomy, and repulsive because the skin was stretched too tight over those huge bones, the invented, sensitive, and unhappy monster duly went on to unleash mayhem on the world at large—and, finally, with appropriate literary even-handedness, upon the over-ambitious doctor too.

Designer humans are not (quite) with us yet. But designer animals have been with us for some time. Charles Darwin, in writing *On the Origin of Species*, was famously dismissive of the 'paltry' geological record as evidence for his new theory of 'descent with modification'. But he noted how, in natural designs, there were relics of the old hidden among the structures of the new. Hence, the whales, thoroughly reshaped from their ancient furry ancestors to become, in effect, very big fish-like animals, nevertheless contained bones that represented the pelvis, femur, and tibia. And he positively enthused about the work of the animal breeders.

Pigeons, and what humans have done to their form, take centre stage in the first chapter of the *Origin*. Darwin said that he had 'kept every breed that he could purchase or obtain', and indeed had a large and thoughtfully designed pigeon loft at Down House, and probably a competent loft manager too; otherwise, he would have got little other work done. He proudly stated that he had been 'permitted to join two of the London Pigeon Clubs', these probably being of the type that kept the membership rate as high as a guinea 'to keep the Spitalfields weaver types out'.[90] He marvelled at the variety of the different breeds, where the differences were not simply of plumage but affected the basic bone structure of the bird. In the skeletons, he said the bones of the face could differ in length and breadth and curvature 'enormously', and that of the lower jaw 'varies in a remarkable manner'. Different breeds had different numbers of ribs, and of vertebrae, and the wishbone could be of a different shape. This was fundamental human-driven re-engineering by selective breeding from the basic design, that he considered to be all derived from the rock dove, with considerable change being possible in just a century.

It was not just pigeons, of course, and Darwin's focus on the work of the animal breeders was to show just how far the fundamental skeletal shape of an animal could be driven from its original, wild structure. Where there have been millennia, not centuries, for humans to reshape an animal then the results can be remarkable. Dogs have been domesticated for at least 14 700 years, since the times of hunter-gathering at the tail-end of an Ice Age. From the original wolf-like ancestor there now have evolved—to give just a few examples—chihuahuas, St Bernards, dachshunds, terriers, greyhounds, miniature Schnauzers, and mutts. Give the bones of each of these to a palaeontologist and he or she would probably diagnose a separate species, based on skeletal characteristics alone.

How quickly can the animal breeders work? Dr Frankenstein would have tipped his hat to the animal reconstructionists of the Chicken-of-Tomorrow programme that started in the USA of the early 1950s, aimed at providing bigger, faster-growing chickens. Chickens had been domesticated for a long time—not for as long as the dog, but since about 2500 BC in south and southeast Asia, the wild form being the Red Jungle Fowl of those parts, a lean, fast-running bird that can live for more than 15 years. Domestication was a success, with the bird being spread by migrants and traders to the Near East and Europe of the Roman Empire, and then carried with the colonists to the New World. Scratching around wherever there were humans, and providing both eggs and meat, the domestic chicken seemed always to be there. Over that time, though, it didn't change drastically. Indeed, establishing the earliest dates of domestication is a problem because the bones of the wild and early domestic forms are so similar. Later on, in places like Europe where it was clear that this bird was introduced, there was a good deal of stasis. Archaeologists often find the bones around Roman and Medieval sites, and have kept track of any changes via standard measurements. A common and recognizable bone is the tibiotarsus, the lower leg bone of the chicken, corresponding to the human tibia, but with several bones of the tarsus (foot) fused to it. The tibiotarsus, through a couple of thousand years, did not vary greatly in dimensions, other than becoming a little thicker at the end overall around the Middle Ages.

Things began to change in the late 19th and early 20th centuries, but the changes moved up several gears once the Chicken-of-Tomorrow team got their hands on the post-World War II chicken. In less than 50 years, the tibiotarsus became more massive, doubling in breadth. The whole chicken, from the wiry and scrawny prewar form, was remodelled into the modern broiler bird, which is four to five times heavier. It is a giant—but an enfeebled giant, for it has been bred not just to be huge, but to grow to be ready for the table from hatching in just 6 weeks. Those enormous bones are riddled with deformities and useless—the bird cannot fly and can barely walk, as its hypertrophied breast muscles

cause it to tilt forwards. It can only live its short lifespan due to continuous human intervention (and, even if rescued from the battery farm, will
quickly die anyway). This—the Anthropocene chicken—is now by far
the commonest bird in the world, the standing stock of more than
20 billion dwarfing the numbers of the commonest wild bird (the red-
billed quelea of sub-Saharan Africa, at about 1.5 billion breeding pairs)
and being further multiplied by the rapid turnover of that absurdly short
lifespan. Moreover, people carefully put the bones, after a chicken dinner,
into the trashcan for burial in landfill sites—a contrast with the rapid
scavenging of almost all wild bird bones after death. This tidy human
habit means that the modern broiler chicken will certainly provide some
of the iconic fossil bones of the Anthropocene Epoch.[91]

There will be, too, the bones of the cattle, pigs, sheep, goats, and other
domesticated animals that humans have shape-changed too and that—
together with us humans—now make up 95% or more of the mass of
medium to large vertebrates on land. The smaller vertebrates, by the way,
are harder to count, but are not likely to change this figure much; the
number of perhaps the most abundant of these, the brown rat, seems
to be roughly comparable to the number of humans worldwide, about
7 billion: as they average about a third of a kilogram each compared with
the average human weight of 62 kilograms, the total brown rat biomass
will weigh in at about half a percent of that of total human biomass.
Bones of all the domestic animals are, like those of the chicken, tidied
away into landfill sites by their human predators, often carefully plastic-
wrapped to aid future fossilization.

This explosion of artificially buried bones are, furthermore, dismem-
bered and/or sawn through in a manner quite distinct from that of
ancient bones, which in past geological epochs were torn apart or
crunched by the claws and teeth of predators less ingenious and technically sophisticated than are modern humans. We really have reached a kind
of peak of skeleton production in this respect. Anthony Barnosky of the
Stanford University has calculated that the total mass of land vertebrates
is now on the order of ten times what it used to be before humans took

control of the food web on land and—crucially—turbocharged it by adding geologically unprecedented inputs of phosphorus (taken from certain, rare rock layers in the ground) and nitrogen (extracted from the atmosphere in the energy-intensive Haber–Bosch process that converts inert nitrogen into usefully reactive ammonia for fertilizers). The crop plants that we now grow so abundantly are mostly fed to our few species of hyper-abundant captive animals before these are fed to us, in this remarkably expanded and modified new global chain of skeletons.

All of this was essentially achieved with just selective breeding, a millennia-old custom, boosted by fertilizer use. Today, the use of the Frankenstein label is mostly with respect to the new technology that has appeared, to remodel biology yet further: genetic engineering, where the animal is altered by direct manipulation of its genes. It is early—and controversial—days so far, and this kind of engineering is mainly directed at cryptic qualities—at least as far as the skeleton is concerned. There are pigs that are modified to excrete less phosphorus, and cows modified to produce low-lactose milk, and even goats that have had silkworm genes inserted into them, in the hope that silk can be woven from an extract of their milk. Things are moving quickly. We suspect that new forms of skeleton, designed in the genetic engineers' labs and putting the Anthropocene chicken in the shade, are not far away.

Of course, one does not need a genetic engineer to modify a skeleton. A real engineer will do.

Augmented Bones

When human skulls from the Neolithic culture are uncovered by archaeologists, as many as one in ten can show a neat round hole in the top of the skull, many with the edges of the bone healed, showing that the owner had survived the hole-making process. This is the result of trepanning, a widely practised operation across both the Old World and the New World, from Stone Age until Medieval times, where a neat hole is made into the skull by either drilling or scraping the bone away. Why was it done? Perhaps, to release evil spirits, or treat recurring headaches. It was a surprisingly widespread practice. For many millennia too, humans have also set bones, or carried out impromptu amputations, following accident or battle—and their skeletons suggest that at least some of these operations were successful. There are also cultural costumes—making the 'lotus feet' of women in China, for instance, a practice that spread through society from the 10th to the 19th centuries. Here, young girls had their feet bound, and toes systematically broken, to achieve the socially desired shape. These were not miniaturized perfect feet, but something more akin to a club foot with permanently deformed bone structure. This practice was agonizing in the making, and painful and difficult to walk on ever after; it was only stopped, by systematic central government pressure, in the 20th century. Or the 'giraffe necks' of the Kayah tribe of Burma, created by inserting successive copper brass rings on to the necks of female children as they grow, the column of rings then kept throughout life. These do not involve great pain as with the lotus feet, and it is not so much the neck that is stretched, as the rib cage beneath that is compressed. Humans have modified their skeletons for a long time.

And added to them, too. Pirate stories, for some reason, have advertised such augmentation most vividly, though in part mendaciously. J.M. Barrie, in creating the eponymous Captain Hook, for instance, spread through popular culture a vision of both pirates and prosthetics that did not exist in reality. Pirates, true, were prone to losing limbs—but the ship's doctor was also usually the ship's cook, presumably because of their everyday practice with carving knife and chopping board. When their patients did not die of gangrene, their remaining stumps would have probably been too painful and mutilated to insert a useable hook, either by surgery or by some complicated kind of harness. An artificial leg of wood, though, seems to have been a more realistic project, and the genuinely fearsome 16th-century French privateer Francis Le Clerc seems to have served as a model for his many descendants in storybook and film (and, occasionally, in real life). Losing one leg during an act of piracy, he had it replaced with a wooden leg and went from strength to piratical strength, being known as 'Peg Leg', 'Jambe de Bois', and 'Pata de Palo' by the awed crew members of the English, French, and Spanish vessels that he attacked without fear or favour. Among his exploits, he devastated the then capital of Cuba, Santiago de Cuba, to such an extent that the city never recovered, and Havana was to become the new capital. His end was appropriate. Refused a pension by Queen Elizabeth I of England (having requested it for his plundering of French shipping) he went on to die with his boots on—or with at least one boot on—in attacking Spanish treasure ships.

Things have moved on since then—and are still moving, ever faster. Augmented skeletons are now safer, much more commonplace, much less melodramatic—and much more *integrated*. When somebody's natural ball-and-socket hip joint wears out today because of old age or disease, part of the upper leg bone can be cut out, to be replaced by a titanium alloy (the ball can be made of alloys of chromium, cobalt, and molybdenum), while the hip socket joint can be removed and replaced by metal, ceramic, or plastic, the whole being fixed to the remaining bones by acrylic cement. Today, some 2.5 million Americans have such artificial hips, about 0.8% of the US population. An infirm knee can be replaced by a cobalt, chromium, and plastic replacement: about 4 million Americans have these, about 1.5% of the population. A broken leg can be pinned by metal, or a broken skull augmented by metal to replace crushed bone. Most commonly by far, a tooth ravaged by sugar in the modern diet can be filled with amalgam, an alloy of mercury, silver, tin, and copper. In the UK, more than 80% of people have at least one tooth filling, and among those, the average is seven fillings per mouth. Most modern human skeletons are re-engineered in one way or another—to permanent, and indeed future palaeontological, effect.

The range of augmentation is now widening rapidly, as the ever-accelerating technological revolution expands to include the human body. Metal bone implants for growing children can now be made bionic, incorporating motors guided by electronic signals, to extend in length inside the body as the children grow. Bionic arms now begin to include moveable fingers with motors, linked to electrodes that respond to signals from muscles in the remaining natural limb. Such augmentation can trump biology—the wrist on such a bionic hand can easily be made to rotate through 360 degrees. The bioengineers who are now speeding the evolution of these devices already talk of being able to replace 50% of a human body by manufactured parts, and aim to link artificial bones with artificial organs and skin into functional systems. How much, the ethicists who these days are part of such design teams ask themselves, of the human body can be replaced before we can no longer say that the result is still human?

It is a good question, because this kind of internal evolution is now colliding with an external evolution that has been taking place among humans for many thousands of years. The ancient Greeks and Romans were among many who appreciated the value of a good homemade exoskeleton.

Do-It-Yourself Skeletons

It is not quite true that only humans can build and modify skeletons. Those skilful protozoans, the agglutinated foraminifera, carefully select size-graded sand grains and shell fragments to build their own combination of exterior armour and living chamber. The caddis fly larva does much the same, as do, in their own way, the remarkable graptolites, both living and dead.

As well as adding to the permanent skeletal mass, humans have been adept at devising temporary additions, from very early days. Creatures without anything special in the way of teeth and claws, after all, needed to be able to improvise, and to improvise quickly and well, to survive. From sharpened flints and antlers, to swords, lances, and arrowheads, the offensive capability developed as a constant of virtually every human society. And because much of the offence came to be directed towards other human social groupings, there was a need to improvise temporary defensive exoskeletons too. Shields, helmets, and armour duly emerged in an arms race similar to that started by that first shell-bearer, *Cloudina*, and its unknown assailant some 550 million years ago.

The arms race created its own myths. That of the Battle of Agincourt in 1415 (Figure 41) was gilded by Shakespeare's narrative powers: that famous day where the arrows of a handful of English archers cut through the suits of armour of the onrushing French aristocracy. It wasn't, though, quite like that quintessentially patriotic account of valour and techno-logical advance. Henry V's 5000 archers did indeed unleash a downpour, probably over 100 000 in a few minutes, of the special 'bodkin' arrows designed to penetrate plate armour, as the 8000 armoured French knights rode on their armoured horses or ran across the 250 metres or so that separated the two lines. In such a deadly hail, not one of these knights should have been left alive. But the defensive armour did its work, the English arrows mostly bouncing off French steel plate, and the French

Figure 41. The massed ranks of the 'exoskeleton'-clad French and English soldiers at the battle of Agincourt, 25 October 1415.

army reached the English lines. But the armour was heavy and the ground underfoot, a rain-sodden wheat field, quickly became a quagmire. The exhausted, now almost immobile aristocrats in their heavy metal casing (greedy for anticipated ransom money, they had pushed the professional French soldiers out of the way to be first in the front line) were cut down not with the longbow, but, as the historical writer Bernard Cornwell put it, 'by all the ghastly paraphernalia of medieval hand-to-hand fighting'—weighted metal hammers, poleaxes, mauls—the horror increasing as Henry V, fearing another attack, ordered all the newly captured French slaughtered. In the struggle for survival, the arms race does not always follow a straightforward course, and it is not only nature that is red in tooth and claw.

Our home-made exoskeletons—and tooth and claw substitutes—have developed apace since then, as Kevlar bullet-proof vests evolve to keep up with advances in bullet technology, tanks vie with artillery shells, and, yet more remotely from the frail human body, satellite-borne lasers lie in wait for nuclear-tipped missiles. Not quite all of these inventions, though, are in the service of the military forces of our modern tribes.

Much like a hermit crab exploiting a discarded gastropod shell—or these days, sometimes a discarded plastic bottle top—for locomotion with added shelter and protection, many of us now climb into our cars each day to commute to work and back. The carefully designed steel panels and bulkheads of our automobiles are not just there to keep the traveller sheltered from rain and wind, but to give protection from other such high-speed wheeled exoskeletons, or from immobile walls or trees, that might be encountered at speed. This is now a new normality for the human species, and these forms of metal cladding can be modified to fly us far through the air and even into outer space, where the astronaut can also use the closer-fitting and more flexible exoskeleton of a space suit, to better explore this inhospitable, airless, and almost endless new realm.

New types of personal hardware are also evolving closer to home, in peace and in war. Newspapers report that the defence industry is preparing 'wearable robots' to create platoons of soldiers with increased load-carrying capacities. Biomedical engineers have already devised similar bionic external frameworks to allow people who have lost the use of their legs to walk once more. This is futuristic stuff—of a future that already seems to be here. And, as it is the same technology at the heart of artificial exoskeletons as underpins artificial endoskeletons, the boundaries of what is outer and what is inner seem set to blur as this dizzying progress proceeds. One might say that the important thing about staying human, as our physical framework is reshaped by our own ingenuity, is the human mind and—in some kind of meaning—spirit.

But will our super-strength, augmented humans be guided by super-strength, augmented minds too? As the bioengineers create new kinds of artificial skeleton, and artificial organs and skin too, the computer scientists are making giant strides with artificial intelligence and autonomous systems. One wonders quite where this will take us, and what will become of the world that we live on. Perhaps it is time to take leave of this Earth, seemingly evolving in the geological blink of an eye to some new state, and see what kind of skeletons one might expect on other planets.

10

SKELETONS ON ALIEN PLANETS

The Curiosity Rover is a wonderful, monstrous machine (Figure 42). It is *big*—the best part of 3 metres long, 3 metres wide, and over 2 metres tall. It weighs very nearly 900 kilograms. Powered by 4.8 kilograms of plutonium, it trundles on its six wheels across the Martian landscape day and night, its fantastical array of cameras and microscopes permanently alert to changes in the rock formations that it passes. Given artificial intelligence by its makers at NASA, if it 'thinks' that a rock

Figure 42. The Curiosity Rover on Mars, a human-made skeleton boldly going where no human has gone before.

looks interesting, it can vaporize part of it with a laser, read the chemical signature with infrared light, and swing its single robotic arm across to interrogate it by X-ray and microscope. These super-exploratory powers are aimed at answering many questions about how Mars formed, and how it has changed over time. But really, truly, at heart—there is one big question there, which drove the multibillion dollar construction of this monster of scientific analysis. It is all about life. Was Mars, in its earlier, milder, wetter days, a living planet? Curiosity is looking for a skeleton.

It is not very likely to find a skull, or limb bone, or coral fragment, or shell. That is a shame for those with a fondness for melodrama, given that alien skeletons are one of the classic elements of science fiction. This fine tradition was most vividly demonstrated by the aliens of the *Alien* films, which are the epitome of evil, and as such it seems perfectly appropriate that they are extremely, chillingly, *skeletal* too. It seemed perfectly appropriate also that Ripley, the human heroine, should don a robotic exoskeleton suit in the final showdown with the alien queen, while the misunderstood but faithful android Bishop, though torn apart early in the confrontation, could still lend a metal hand in battle. Whether human, alien, or android, these all have *advanced* skeletons. The history of our planet, and what we know of the structure of others, suggest that we should rather expect something quite different, as we begin to explore worlds beyond ours.

Earth is a planet that somehow has turned out to be ideally suited to maintaining life continuously for something on the order of 4 billion years. But the kind of skeletons that one saw in *Aliens*—if we allow the aliens to have roughly dinosaurian body plans—would only have appeared in the last 5% or so of that time. For the 10% or so before that, external carapaces mostly ruled the roost. Some of these organisms could, as we have seen, be impressive, such of that 2-metre-long proto-millipede *Arthropleura* of Carboniferous times—but mostly they were on a different and, usually, smaller body plan.

This, one suspects, is a vision that will encompass most life-bearing planets in the cosmos. On Earth, multicellular animals and large, sophisticated skeletons only arose a little over half a billion years ago, after more than 3 billion years of continuous evolution of a microbial biosphere. We must recall also that such organisms, no matter how well armoured, only have another billion years of existence at most on Earth, before the inexorable brightening of the Sun boils away both oceans and atmosphere. On the evidence of Earth, it seems possible that quite a few planets could get to the stage of microbial life, but that either they will never arrive at the more difficult stages of eukaryotic cells and multicellularity or, if they do, those more complex experiments in life will perish in some planetary vicissitude or other. Some 252 million years ago, perhaps 19 out of 20 multicellular species perished in the wake of an episode of savage prolonged volcanism at the Permian–Triassic boundary. The Earth seems to have come close, then, to losing the last, twentieth part of those species—and being knocked back to a wholly microbial, Precambrian-style global ecosystem.

Earth, therefore, may be rare as a planet in harbouring life-friendly conditions, for so long, to finally enable such a difficult biological trick as multicellularity to be accomplished. By comparison, Mars may have had just a few hundred million years for life to appear and evolve. Even Venus, now a hellish and quite sterile ultragreenhouse world, might have had an early habitable phase. Such planets, with their brief potential windows for life, are unlikely to have gone beyond the microbial world. So an exobiologist—or exopalaeontologist—exploring other planets might do well not to have a mental focus on ribcages and skulls, or bizarre carapaces. A more sensible focus might be on finding a microbial skeleton.

Microbial skeletons, of a kind, do exist—and they can be big enough to see, easily, with a human eye or by Curiosity's electronic eyes (of which it has at least 17). Microbial skeletons abounded in some parts of the ancient Precambrian world, and some even survive today. But their study has been fraught with blind alleys, wrong turnings, serendipitous discoveries, mistaken interpretations. Even now, after more than a century of study, including by some of the most eminent geologists and palaeontologists to have wielded a hammer, they are not always simple, nor straight-forward, to recognize.

Hence, as prime targets for the intrepid explorers of Mars and more distant planets, it is worth recalling the cautionary tale of the discovery of microbial skeletons on Earth, for this may provide essential preparation for some of the conundrums to come, as the hunt for evidence of extraterrestrial life, present or past, intensifies. There may well be more blind alleys and wrong turnings in store, to torment the lives of the new generation of exopalaeontologists, and so this new generation of scien-tists might do well to consider the mysteries that their ancestors grappled with. To understand future science on other planets, we can start by staying at home, and going back to the deep past, to when the Earth itself was—compared to its current state—a thoroughly alien planet.

Eozoön, or the Proto-skeleton That Wasn't

William Logan was a man who grew illustrious. Canadian born, but educated in Edinburgh, Scotland, he made himself well known enough as a geologist working in the South Wales coalfield to be headhunted to create, and become the Director of, the Geological Survey of Canada. In the course of a glowing career, he was awarded 27 medals, a knighthood, had a mountain named after him (Mount Logan, the highest in Canada), the edge of a line of mountains (Logan's Line, which marks the western edge of the Appalachians) and a fossil—as regards *both* the genus and species (the graptolite *Loganograptus logani*). He was an experienced geolo-gist who had seen lots of rock. And he was well aware of the phenomenon that had so baffled Charles Darwin—the presence of fossil skeletons in Cambrian and younger strata, often so plentiful that they could crowd the rocks, and their puzzling absence from the huge stretches of older rocks, which they knew must represent vast spans of time.

So when a rock specimen was brought to him by one of the Survey's collectors in 1858, from rock exposures of some of these very ancient

Precambrian rocks by the Ottawa River, it rang a bell with him—and made him think. He had seen similar rock specimens near Perth, Ontario. They both had a striking concentric pattern, one made of alternating layers of the minerals calcite and pyroxene, and the other of dolomite and (but of what else?) loganite. The pattern looked like a fossil, and yet the conventional wisdom of the day said that the rocks were too old for it to be such a thing. The rocks had also been intensely deformed and recrystallized by tectonic forces, a process that usually obliterates fossils. Nevertheless, Logan thought that these patterns might really be fossils of the earliest life on Earth, which he called *Eozoön canadense*.[92] He showed them at meetings in both America and England—but found more sceptics than believers.

Acceptance only grew when he sent specimens, which had been cut into 'thin sections'—slices cut and ground to a thousandth of an inch in thickness for analysis by microscope—to another established and well-respected geologist, J. William Dawson, then Principal of McGill University in Montreal. Dawson's credentials were equally secure—he was a protégé of the yet more famous Charles Lyell, the English scholar who more than anyone else founded modern geology. Dawson examined the thin sections and declared that they represented the shells of monstrous foraminifers—and so that they must indeed be organic. Foraminifers today are organisms that build exquisite skeletons of calcite—but they are generally tiny (they are single-celled, and relatives of amoebae), rarely exceeding a millimetre in size. These ancient versions were giants. Dawson grew lyrical. This 'remarkable fossil', he said, 'will be one of the brightest gems in the scientific crown of the Geological Survey of Canada'.

The bandwagon for this avowedly earliest of all earthly skeletons began to move. The palaeontologist James Hall (coiner of the name *Loganograptus*) concurred with Dawson's identification, and called *Eozoön* 'the greatest discovery in geology for half a century at least'. Both Charles Darwin and his 'bulldog' of an academic protector, Thomas Huxley, agreed that these seemed to be true fossils, with Darwin noting the discovery in his 1866 edition of the *Origin of Species*. It became one of the highlights of a presidential address given by Lyell himself, to the

British Association for the Advancement of Science. Not all the scepticism went away, though—and a broadside against *Eozoön* came from an unexpected quarter.

Two Irish mineralogists, William King and Thomas Rowney, based at the Queen's College in Galway, published their counterblast to Logan and Dawson in 1866, in the pages of the *Quarterly Journal of the Geological Society of London*. Although originally 'zealous advocates of the organic origin' of *Eozoön*, they now—'after prolonged investigation'—begged to differ. The structures identified as chambers of fossil foraminifer shells were, they said, preserved in asbestos-like minerals in profoundly metamorphosed rocks: these were purely mineral growths, and not fossils at all.

Their paper started a controversy—initially polite and good humoured, but later growing increasing rancorous—that was to go on for three decades. The opposing schools of eozoonists and anti-eozoonists were to question each other's evidence and logic, and competence too, while fresh specimens and counter-specimens were sought and found, and brought into the fray. Logan seems to have taken little part in these further discussions. Laden with honours, he eventually retired to a quiet village in Pembrokeshire, south Wales. Dawson, though, was to carry on taking up the cudgels for *Eozoön* until the end of his long life—although he was dismayed by Darwin's mention of his cherished fossil, for he was a fervent antievolutionist.

By the early 20th century, there was common agreement. *Eozoön* was a pseudofossil, or false fossil: a complex, purely inorganic segregation of minerals formed under specific conditions of temperature and pressure and rock type, and so not an organic remain at all. The length of the debate—and the number of competent, indeed excellent, geologists on both sides—was a measure of how easily fossil skeletons can be mimicked by purely physical and chemical processes—and how difficult it can be to distinguish one from the other. *Eozoön* was not the only pseudofossil mistaken for, and named as, a real fossil in those ancient rocks. There were other such impostors—*Archaeospherina*, *Manchuriophycus*, *Rhysonetron*, *Brooksella*, and more. Most have been now debunked, or remain as enigmatic 'dubiofossils'.

Micro-skeletons on Mars?

So, even with abundant rock evidence that could be examined minutely by experienced geologists with the best microscopes of the day, it was all too easy to make profound misinterpretations. One here is reminded of the announcement of the discovery of fossilized bacteria in the Martian meteorite ALH840001, as announced on the White House lawn by President Clinton in 1996. This startling statement sparked intense discussions, for and against this interpretation, much as *Eozoön* had done in Victorian times. The outcome was similar too—these tiny structures were probably not bacteria, but simply microscopic mineral growths.

Scientific life amid the alien worlds of a distant planet, or the distant and alien times of our own planet, is thus full of pitfalls. But, as Curiosity trundles its multi-billion-dollar pathway across the Martian surface, it is worth remembering that ALH84001 is only one of more than a hundred Martian meteorites so far recognized, and that many more certainly lie in the main meteorite hunting grounds of Earth: within the snow and ice of Antarctica and in desert regions. Most of the currently known Martian meteorites are igneous in origin—even the notorious ALH84001—and so are generally unpromising rock types in which to look for fossils. One or two are breccias, which offer more possibilities—though brecciation occurs not just by normal surface erosion but also as the incoming bolides blast rock fragments off Mars's surface. Perhaps one needs to look harder on Earth for evidence of life on Mars. Martian meteorites of some sandstone or mudrock may be the real prize—and this may not be an impossible prospect, given the wide extent of sedimentary rocks on Mars that would also have been pummelled from space.

A gram or two of mudrock on Earth can contain tens of thousands of mini-skeletons, invisible to the naked human eye, such as the acritarchs. These have been forming on Earth for over 3 billion years, well before larger, more obvious shells and bones evolved. Just perhaps, some similar kind of mini-skeleton might have evolved on Mars in its first billion years before its surface became dry, frozen, and steeped in bactericidal chemicals such as perchlorates.[93] Curiosity will be hard pressed to spot such tiny skeletons, even if they are present in the stratal vistas it is driving over. But if there were even a small chip of such rock on Earth, a skilled palaeontologist would have a fighting chance of unearthing such microscopic relics.

The ancient Precambrian strata of Earth do contain large structures that really are the result of primitive life, and that can even be considered as primitive skeletons, and that—if they are present—can be readily studied by Curiosity's 17 eyes. This almost unimaginably long span of time was the heyday of the stromatolites, when microorganisms began to crystallize minerals around them to create particular kinds of rock textures that one might regard as a primordial skeleton. Perhaps it is *the* primordial skeleton, for it might represent some kind of interplanetary constant, rather more so than do the sophisticated and baroque skeletons that are the hallmark of the current Phanerozoic Eon on Earth. Stromatolites represent one kind of primitive structure that is now one of the main items on the wish list of the tireless Curiosity Rover. But study of the earthly stromatolites has not been at all straightforward—and that of any Martian candidates may not turn out to be any easier.

Living Rocks?

Some decades before William Logan and J. William Dawson set off the wild goose chase that was *Eozoön*, the geologist John Steele had found rocks with rounded bundles of fine layers, looking a little like cabbage heads, in Cambrian rocks of New York State.[94] He called them 'calcareous concretions', which implied that he thought them inorganic structures, produced by some kind of subterranean chemistry. These structures were later, in 1865, named *Cryptozoon proliferum* by the same James Hall who had invented *Loganograptus logani*, and who had blessed Logan's and

Dawson's interpretation of *Eozoön*. Enthused by the seeming break-through in recognizing ancient life, Hall reclassified Steele's 'concretions' as biological, not chemical, declaring them to be some kind of early animal skeleton. Similar structures were later found in other Precambrian rocks, and named *Archaeozoon*. By the late 19th century, various layered structures had been found in Precambrian and Cambrian rocks, and their interpretation revolved around the most basic of questions: animal or mineral? So, it was fitting that the only other available option—vegetable—would then enter the field of play.

It was Ernst Louis Kalkowsky[95] who provided this suggestion, working in new territory, the relatively young and unaltered rocks of Triassic rocks of Europe. In the late 19th century, Kalkowsky was a noted geologist in Saxony, director of the State Museum of Mineralogy and Geology at Dresden. There is a photograph of him at a geological meeting in Dresden in 1895: all present (except for the pianist) are cheerfully and theatrically holding steins of beer towards the camera. With this, and the fine assembly of beards and moustaches on view, it resembles several multiples of the Marx brothers at their most high spirited. Kalkowsky was clearly operating within a thriving and energetic geological community. Kalkowsky's position at the museum reflected his own interests—and prejudices. He regarded himself as a mineralogist and geologist, and rather disdainfully considered palaeontology to be an 'ancillary subject', more related to biology than to geology. So it is apt, in the curious way that the Earth sciences often operate, that his greatest and most memorable insights were palaeontological in nature.

Early in the 20th century, while working on the early Triassic rocks of Lower Saxony, he noticed curved layers of finely laminated limestone, often forming dome-shaped masses a few centimetres high. They looked rather like the North American *Cryptozoon*, though Kalkowsky was prob-ably unaware of this term, given his general disparagement of palaeon-tology. He termed these layered dome-shaped structures *stromatoliths* ('layered stones'), and said that they had formed by the action of 'simple types of plant-like organisms'—that is, in effect, they were microbially generated structures formed of successive growing layers. Kalkowsky's new term, which soon evolved into 'stromatolites', became widely adopted. But his interpretation quickly ran into opposition from geologists who said that these were simply layered chemical precipitates. In essence it was a repetition of the *Eozoön* debate with a central question at its heart: what is biological, and what is chemical, in such structures? When Kalkowsky died in 1938, the balance of opinion was still that they were chemical in origin.

More evidence was to come—and this time from the present, which once more was to show itself as a key to the ancient past.

Living Stromatolites

Charles Doolittle Walcott, who was discoverer of the Cambrian fossil bonanza of the Burgess Shale in British Columbia, was part of a lineage that, in at least one way, was *extremely* conservative. His father was also called Charles Doolittle Walcott, and he carried on the tradition by giving exactly the same name to one of his own sons (though, thankfully, not to all of them). Stephen Jay Gould, who portrayed Walcott in his book *Wonderful Life*, considered that he embodied conventional beliefs. But, as regards stromatolites, Walcott's work showed much imagination and insight. By some quirk of coincidence, Walcott's career intersected with James Hall's— he was once Hall's assistant—and it paralleled William Logan's (he became head of the United States Geological Survey, before taking the reins at the Smithsonian Institution). Walcott also had a mountain named after him (although admittedly just one peak, part of Mount Burgess).

Walcott, in his geological fieldwork, intermittently encountered bulbous layered rock structures that he referred to Hall's *Cryptozoon*, before coining a term for another, less markedly dome-like variety of this distinctively layered rock, which he termed *Collenia* (after a rancher in the Big Belt Mountains of Montana). Independently of Kalkowsky, in 1914, he suggested that these structures were the result of the activity of

'Cyanophyceae', what we now call cyanobacteria.[96] He went further than Kalkowsky, though, in suggesting a modern analogue. He pointed to examples of mound-like structures coated in sticky microbial mats that could be glimpsed, just below the surface of shallow water in lakes in limestone terrain. These structures, called calcareous tufas, he suggested as modern versions of those ancient Precambrian structures.

Walcott's ideas, like those of Kalkowsky, got a mixed reception. Not all researchers were convinced by the tufa analogue. One problem was that modern tufas, while definitely associated with the cyanobacteria covering them, are much more porous than the ancient stromatolites, and their layering is often poorly developed—so there was still an impasse. The next breakthrough was to come from the sea.

One of the nice things about working on limestone rocks and fossils is that, to explore modern examples, some of the best places are hot, sunny, picturesque, and by the seaside. From the 1930s, geologists had been travelling to the shallow-water paradises of place like the Bahamas and Florida, to see modern limestone textures and structures forming literally before their eyes. Some of these geologists also had their eyes open for any clues that might help solve the stromatolite mystery. Plenty of microbial mats were found, but no close analogues of those ancient Precambrian structures.

The real breakthrough came from the other side of the world—albeit also in a swelteringly hot coastal region. At the most westerly tip of Australia lies Shark Bay, a large stretch of water partly cut off from the Pacific Ocean by a line of sandbars. When the French explorer Henri-Louis de Saulces de Freycinet encountered it at the beginning of the 19th century, he thought (wrongly) that some of the sandbars wholly prevented access to what was otherwise an inviting harbour, and so named it 'Havre Inutile' (i.e. 'Unusable Harbour'). Hence—following a few liberties taken in the translation—the town now glories in the name of Useless Loop.

The inhabitants of Useless Loop make a living by winning salt from the sea, allowing the baking sun to quickly evaporate water from large saltpans. This strongly evaporitic regime, and the highly saline water in the

Figure 43. Living stromatolites at Shark Bay in Western Australia.

bay, is a factor in preserving what later became Shark Bay's chief claim to fame, as the home of some living and impressive stromatolites (Figure 43).

These stromatolites were described in the mid-1960s by another geological Logan: Brian Logan (not a relative, as far as we know, of William Logan). He directly compared these forms with both *Cryptozoon* and *Collenia*[97] of the Precambrian. The modern structures are abundant in the intertidal zone of the lake, looking like nothing so much as scattered forests of inverted elephants' feet: low columns, a little expanded at the tops, which are flat and which reach the high tide level.

Examined more closely, the tops of the columns are covered by a thick, rather rubbery microbial mat, made of a mesh of microscopically fine cyanobacterial filaments a few millimetres thick. Beneath the mat, the columns are formed from curved layers made of the sandy sediment in the bay. Logan suggested that the microbial mat secreted sticky organic substances that initially trap the sand to build up sediment layers. As soon as one layer is built up, another mat quickly forms on top to trap yet more sediment. The microbial mats stimulate natural cementation of the grains by calcium carbonate, and so these primitive microbial skeletons are quickly converted into newly formed rock.

Why Shark Bay? And why are these structures so rare elsewhere in the modern world? Logan emphasized the setting: a semienclosed bay set in a hot evaporative climate. The resulting high salinities discourage animals such as snails and crustaceans, which would otherwise have eaten their way through the microbial mats. In the Precambrian, of course, there were no snails or crustaceans, and so the microbial mats—and stromatolites—could flourish and build undisturbed widely across the sea floor. It is the primitive beginning of large-scale skeleton building—a stage in which the early Earth was stuck for some 3 billion years.

Just perhaps, Mars might have got to that stage too.

Martian Stromatolites?

It was the Spirit Rover, a direct ancestor of the Curiosity, that detected hints of stromatolite-like structures on Mars. These were not in carbonate rocks (which are rare on Mars), but in silica, of the hydrated, opaline kind that forms around hot springs. Photographs taken by the Spirit Rover showed small knobbly structures of this mineral poking out of the soil. They were described as 'cauliflower like' by the space scientists involved—exactly the same analogy that some of their Victorian forebears used to describe ancient earthly stromatolites. Were these, therefore, crude microbial skeletons that show that life existed on an ancient Mars? Even more tantalizingly, these Martian cauliflowers resemble structures that today form around geysers in the hyperarid Atacama desert of Chile—one of the few places on Earth that comes close to replicating some of the conditions of Mars on Earth. Microbes are thought to play a part in forming finger-like protrusions on these Chilean structures, which have been described as 'microstromatolites' by the researchers who studied them. So, is this evidence for genuine Martian skeletons?

'Not so fast' was the general response. Just because something looks biological doesn't mean that it *is* biological. Here, one can reach back from the red planet to consider our own, upon which the stromatolite debate has become more nuanced in recent years. It is not a question of whether biologically produced stromatolites have existed in the past. They have, of course, for we can see that they are present now. The question being asked now is whether all stromatolite-like structures are biological—or whether there are many that are purely chemical in origin. If so, then stromatolite-like structures on Mars and other planetary bodies—where any microbes may well have not played to earthly rules—need to be considered *very* cautiously.

The Earth's oceans, it should be recalled, are full of calcium, carbonate, bicarbonate, and other ions, the dissolved products of the chemical weathering of rocks on land, brought in large amounts to the sea by rivers. Once the oceans are saturated with these ions, they will precipitate out on to the sea floor as layers, whether living organisms are helping the process or not. And there are layered rock structures in other environments on Earth, such as the stalactites and stalagmites in limestone caves that form purely chemically. The same will be true of any kind of chemically saturated fluids in different environments on any planet. So how might layers made chemically be distinguished from those made biologically?

Close examination of ancient stromatolite-like structures on Earth has revealed two kinds of layering.[98] One is of more or less continuous layers made up of finely crystalline (or 'sparry') calcite, and one is of more uneven layering typically made up of cemented carbonate sediment. The former is now thought to be chemical in origin, and the latter biological (though hybrid forms are also known). Using such criteria, the oldest, reasonably convincing biologically produced earthly stromatolites go back to some 3.5 billion years ago.[99] At Strelley Pool in the Pilbara region of Australia, there are sedimentary rocks that have escaped the kind of severe metamorphism that most other Archean rocks have been subject to. The rocks there include a range of stromatolites—some low with gentle slopes, others like egg-crates, and yet others as high, linked cones. This range of shapes cannot be easily explained by pure chemical precipitation, but made sense if different kinds of sticky microbial mats were invoked. They are the oldest of these proto-skeletons on Earth—and formed at a time when liquid water still ran across the surface of Mars.

No fossilized microbes have been found within the Strelley Pool stromatolites. However, there are associated layers of chert in the region: exceptionally fine-grained silica-rich rocks that have long been known to have the capacity to preserve the soft tissue of individual cells. The hunt for such fossils took a course that is now becoming familiar. An early report of filamentous microfossils was later debunked—the filaments were simply cross-sections through fractures that were filled with iron and silica minerals. But more careful work on other layers found convincing examples of microfossils.[100]

All this gives hope to those scientists who are seeking stromatolites on Mars and other planetary bodies. Of course, the rules of thumb that allow distinction of these crude biological skeletons from chemical precipitates may not be transferable beyond Earth. Each planet that may have, or once have had, life will possess its own rules of thumb. Thus, while keeping a sharp eye out for the layered structures that might be exo-stromatolites, it should be recalled that biologically produced layered rocks are only different *in detail* from layers that arise out of purely physical or chemical processes. The devil is always in the detail in such cases—and different planets will have, for sure, different devils.

Diversifying Extraterrestrial Skeletons

One might look on the extraterrestrial skeleton search through another prism. Water is necessary for life as it is currently understood. The Earth is the only body in the Solar System with abundant liquid water at its surface—right now, that is: Mars and Venus might both have had surface oceans in the past.[101] But, as human exploration of the Solar System has proceeded, liquid water oceans have been discovered—some even more voluminous than those of Earth. Europa, Titan, Callisto, Ganymede, Enceladus, Triton, Mimas—all of these moons of other Solar System planets have proven or suspected oceans, although hidden under a thick carapace of ice.

So, here let us consider a main ingredient of skeletons—some kind of dissolved substance that can more or less easily crystallize out of solution to provide the rigid mineral structure for a skeleton. On Earth, the prime materials are calcium carbonate, silica, phosphate, and complex organic substances. On other planets, there may be a different array of such handy substances. Mars, for instance—both in the long-gone days when it had some kind of hydrosphere and subsequently—has and likely had some kind of deficit of carbonate material, perhaps because it was so easy on that planet for a prime ingredient, carbon dioxide, to be whisked off to outer space by the solar wind. Silica, though, was a major ion in its diminishing seas and lakes. So too was sulphur, as sulphate. On Earth, sulphate is used by microbes as an energy provider, to be reduced down to sulphide, which can then link with iron to produce pyrite. So might we suspect gleaming skeletons of fool's gold on Mars?—akin to the pyrite carapaces that earthly fossils can acquire after death, or the iron-sulphide rimmed skeletons that are currently formed in life by at least one deep-sea gastropod?

One can imagine other skeleton-forming substances. Ice might come into play on planets a little colder than ours, or on those Solar System moons. Or salts of magnesium, potassium, sodium, and selenium. The hydrocarbons may be a constant in skeleton production. Think of Titan, where rains of methane and ethane fall to make up rivers, lakes, and seas of hydrocarbons, upon a landscape of rock-like ice. Those hydrocarbons commonly polymerize to make up hazes and particles of more complex, high-molecular-weight compounds, with some of these then being blown by the nitrogen winds into shifting dunes. Perhaps this environment, at minus 180°C, is not the place to incubate life (as we know it, at least). But, beneath that icy carapace, there is a water ocean that might have dissolved salts from the rocky core of the moon—and where those surface organic compounds might filter down through cracks and fissures to contaminate—and perhaps fertilize—the dark, sunless depths. Here, energy may be derived through physical means such as tidal forces, or chemical means such as liquid–rock reactions, to power some kind of life. And for that life, with those hydrocarbons to hand, to experiment with minuscule acritarch-like skeletons, Titan-style, might be imagined to be part of some exo-ecological system.

What then if—in some of those places—life was eventually to develop the sophisticated command-and-control systems that can cause many cells to live and act together in concert, and diversify to produce special-ized tissues? What might the next skeletons be?

Among the possibilities of some kind of universal next-stage skeleton might be coral-like structures, building up from some kind of ocean floor. After all, on Earth this is a pattern that has been taken up not only by three very different kinds of corals, but also by sponges, stro-matoporoids, archaeocyathids, bivalves, brachiopods, and bryozoans. And where mobility and protection is needed, the bivalved shell has been a hardy perennial, not just with the bivalve molluscs, but also with brachiopods, conchostracans, and ostracods. And when it comes to longevity and survivability, forget the giant clam—it is the small things that inherit the Earth. The rice-grain-sized ostracods have come through pretty well every one of Earth's mass extinction events: if not entirely unscathed, then at least considerably less scathed than were most other groups of organisms. The meiofauna have probably been a constant too. So an exobiologist or exopalaeontologist, on landing from their inter-stellar starship, may be well advised to have a simple but powerful hand lens with them at all times.

The type of skeletal organism to fear, though, may be the one that has arisen just once, the product of just one of the billion or so species that have inhabited this planet. These are the technological skeletons that arose, geologically, only minutes ago, and which have since evolved to take over and transform the planet—and are still evolving, ever faster. The story of 'us' and of our machines might, admittedly, end soon by any one of dozens of possible sources of catastrophe. Or we and machines both might negotiate the next few perilous centuries to take up a more permanent residence on Earth, and be seeking to colonize other planets. In such distant and hostile regions, skeletons made of metal and silicon are likely to be more reliable and durable than is our familiar but primitive model, made of phosphate bones with soft surrounding tissue.

Presumably, some other planets, somewhere in the Universe, might have reached such a turning point too, an abrupt bifurcation in history that can end in tragedy or transformation. Perhaps a mature biological history can only really start in discarding the bones of one's ancestors, and designing skeletons that can support life among the stars.

NOTES

Chapter 1

1. Hua et al. 2005.
2. Some sea anemones, though, have evolved to be able to burrow through sediment.
3. Babcock et al. 2014.
4. Hua et al. 2003.
5. Chen et al. 2008; Cai et al. 2015.
6. Lee et al. 2013.
7. Vartanyan et al. 1993.
8. We give the full story in Chapter 6 of *Ocean Worlds* (Zalasiewicz and Williams 2014).
9. E.g. Gabbott et al. 2008.
10. Caron and Jackson 2008; Vannier 2007.
11. Moysiuk et al. 2017.
12. Welland 2009.
13. Rundell and Leander 2010.
14. Fortey et al. 1997.
15. Harvey and Butterfield 2017.

Chapter 2

16. Hou et al. 2014.
17. Butterfield and Harvey 2012.
18. Knell and Fortey 2005. See also Richard Fortey's excellent book *Trilobite!* (Fortey 2000).
19. Campbell 1975.
20. Hou et al. 2017.
21. Selden and Read 2008.
22. Choo et al. 2014.
23. Dunlop 1996.
24. Yao et al. 2010.
25. Yoon 1995.
26. Vermeij 1995.
27. Daniel et al. 1997.

Chapter 3

28. To a first approximation, it is the planet of the microbes, and has been from some four billion years ago.
29. In this role, we recommend readers to the Archy (the cockroach) that befriended Mehitabel, the scandalous cat, and wrote the *vers libre* poems brought to us by Don Marquis (a human).
30. Sharp-eyed viewers of Hayao Miyazaki's *Castle in the Sky* may catch a glimpse of them, tranquilly swimming in the castle moat.
31. We describe the remarkable discovery of this fossil locality in *Ocean Worlds* (Zalasiewicz and Williams 2014).
32. Agassiz immortalized the discoverer of these bizarre fish, Hugh Miller, by naming as *Pterichthys milleri* one of the species that the talented stonemason had found.
33. One of the team, Neil Shubin, subsequently wrote the classic book *Your Inner Fish*, part of which recounts the discovery (Shubin 2009).
34. Paton et al. 1999.
35. The film with Hollywood actor Victor Mature, and not the remake with Raquel Welch.
36. By Jostein Starrfelt and Lee Hsiang Liow of the University of Oslo (Starrfelt and Liow 2016).
37. Groenewald et al. 2001.
38. Brunet et al. 2002.
39. Villmoare et al. 2015.
40. Ward et al. 2014.
41. Henshilwood et al. 2002.
42. Zamora et al. 2012.
43. On what came to be known as evolution, to be precise. Darwin himself did not use the term, calling it 'descent with modification'.
44. Gahn and Baumiller 2003.

Chapter 4

45. Knauth and Kennedy 2009.
46. Stein et al. 2007.
47. Buffon 2018.
48. Winchester 2001.
49. Davies and Gibling 2010; Davies and Gibling 2013.
50. Williams et al. 2014.
51. Williams 2010.
52. Lateiner 2002.

Chapter 5

53. Including at the Triassic–Jurassic boundary, the mass extinction event that only relatively recently has attracted the kind of study devoted to the Permian–Triassic and Cretaceous–Tertiary: see Pálfy and Kocsis 2014.

Chapter 6

54. We describe the birth and death of this extraordinary concept in our book *Ocean Worlds* (Zalasiewicz and Williams 2014).
55. This seems to be currently the 'superacid' fluorantimonic acid, which is about a million times more acidic than concentrated sulphuric acid.
56. Javaux et al. 2010.
57. It did not permanently put him off travelling. A later expedition to Russia was much better organized, though one observer noted that a significant hazard was the excessive hospitality of the Russian hosts.

Chapter 7

58. Hu et al. 2016.
59. Trewin 2004.
60. Engel and Grimaldi 2004.
61. Norberg 2006.
62. Chapelle and Peck 1999.
63. Xu et al. 2013.
64. Bressan 2014.
65. Hone 2009.
66. Witton 2015.
67. Though we note that the age of the deposits this fossil is sourced from is debatable, and it could be as young as the Early Cretaceous.
68. Balter 2015.
69. Morris 2017.
70. Meng et al. 2006.
71. McCracken et al. 2016.

Chapter 8

72. Morelle 2013.
73. A phrase of the environmental campaigner Rachel Carson. Best known for her book on pesticides *Silent Spring* (Carson 2000), she also wrote fine books about the sea.
74. Hays et al. 1976.

75. The 100 000-year cycles in reality may individually be stretched or squeezed by the influence of the 20 000-year and 40 000-year cycles. The latest one has been stretched in this manner.
76. Mauritian Wildlife Foundation 2017.
77. Wells 1963.
78. Scrutton 1964.
79. Trotter et al. 2008.
80. Schopf 1970.
81. Du Hamel and de Buffon 1737.
82. Her novel on this theme, *The Daughter of Time*, is widely considered one of the best detective stories ever written.
83. Lamb et al. 2014.

Chapter 9

84. The non-avian dinosaurs, that is. The avian dinosaurs—the birds—are still with us.
85. Mayor 2000.
86. Smil 2011.
87. Koch and Barnosky 2006.
88. Barnosky et al. 2011.
89. Dixon 1983.
90. Not Darwin's own words, we hasten to add: but see Ross 2017.
91. As researched by a team led by Carys Bennett and Richard Thomas of Leicester University.

Chapter 10

92. There is much fascinating detail of this story in O'Brien 1970; and Adelmann 2007.
93. Wadsworth and Cockell 2017.
94. Riding 2011.
95. Gehler and Reich 2008.
96. Riding 2011.
97. Logan 1961.
98. Riding 2011.
99. Allwood et al. 2006.
100. Wacey et al. 2011.
101. Possibilities that we discuss in *Ocean Worlds* (Zalasiewicz and Williams 2014).

REFERENCES

Chapter 1

Babcock, L.E., Peng, S., Zhu, M., Xiao, S., and Ahlberg, P. 2014. Proposed reassessment of the Cambrian GSSP. *African Journal of Earth Sciences* **98**, 3–10.

Cai, Y., Xiao, S., Hua, H., and Yuan, X. 2015. New material of the biomineralizing tubular fossil *Sinotubulites* from the late Ediacaran Dengying Formation, South China. *Precambrian Research* **261**, 12–24.

Caron, J.B., and Jackson, D.A. 2008. Paleoecology of the Greater Phyllopod Bed community, Burgess Shale. *Palaeogeography Palaeoclimatology Palaeoecology* **258**, 222–56.

Chen, Z., Bengtson, S., Zhou, C., Hua, H., and Yue, Z. 2008. Tube structure and original composition of *Sinotubulites*: shelly fossils from the late Neoproterozoic in southern Shaanxi, China. *Lethaia* **41**, 37–45.

Fortey, R.A., Briggs, D.E.G., and Wills, M.A. 1997. The Cambrian evolutionary 'explosion' recalibrated. *Bioessays* **19**, 429–34.

Gabbott, S., Zalasiewicz, J., and Collins, D. 2008. Sedimentation of the Phyllopod Bed within the Cambrian Burgess Shale Formation of British Columbia. *Journal of the Geological Society* **165**, 307–18.

Harvey, T.H.P., and Butterfield, N.J. 2017. Exceptionally preserved Cambrian loriciferans and the early animal invasion of the meiobenthos. *Nature Ecology and Evolution* **1**, 0022.

Hua, H., Chen, Z., Yuan, X., Zhang, L., and Xiao, S. 2005. Skeletogenesis and asexual reproduction in the earliest biomineralizing animal *Cloudina*. *Geology* **33**, 277–80.

Hua, H., Pratt, B.R., and Zhang, L-Y. 2003. Borings in *Cloudina* shells: complex predator–prey dynamics in the terminal Neoproterozoic. *Palaios* **18**, 454–9.

Lee, M.S.Y., Soubrier, J., and Edgecombe, G.D. 2013. Rates of phenotypic and genomic evolution during the Cambrian explosion. *Current Biology* **23**, 1889–95.

Moysiuk, J., Smith, M.R., and Caron, J-B. 2017. Hyoliths are Palaeozoic lophophorates. *Nature* **541**, 394–7.

Rundell, R.J., and Leander, B.S. 2010. Masters of miniaturization: Convergent evolution among interstitial eukaryotes. *Bioessays* **32**, 430–7.

Vannier, J. 2007. Early Cambrian origin of complex marine ecosystems. In: Williams, M., Haywood, A.M., Gregory, F.J., and Schmidt, D.N. (eds) *Deep Time Perspectives on Climate Change*. Geological Society.

Vartanyan, S.L., Garutt, V.E., and Sher, A.V. 1993. Holocene dwarf mammoths from Wrangel Island in the Siberian Arctic. *Nature* **362**, 337–40.

Welland, M. 2009. *Sand*. Oxford University Press.

Zalasiewicz, J., and Williams, M. 2014. *Ocean Worlds: The Story of Seas on Earth and Other Planets*. Oxford University Press.

Chapter 2

Butterfield, N.J., and Harvey, T.P.H. 2012. Small carbonaceous fossils (SCFs): A new measure of early Palaeozoic palaeobiology. *Geology* **40**, 71–4.

Campbell, K.S.W. 1975. The functional morphology of *Cryptolithus*. *Fossils and Strata* **4**, 65–86.

Choo, B., Zhu, M., Zhao, W., Jia, L., and Zhu, Y. 2014. The largest Silurian vertebrate and its palaeoeocological implications. *Scientific Reports* **4**, 5242.

Daniel, T.L., Helmuth, B.S., Saunders, W.B., and Ward, P.D. 1997. Septal complexity in ammonoid cephalopods increased mechanical risk and limited depth. *Paleobiology* **23**, 470–81.

Dunlop, J. 1996. A trigonotarbid arachnid from the Upper Silurian of Shropshire. *Palaeontology* **39**, 605–14.

Fortey, R. 2000. *Trilobite!* Harper Collins.

Hou, X., Siveter, D.J., Siveter, D.J., Aldridge, R.J., Cong, P., Gabbott, S.E., et al. 2017. *The Cambrian Fossils of Chengjiang, China: The Flowering of Early Animal Life*. John Wiley and Sons.

Hou, X., Williams, M., Siveter, D.J., Siveter, D.J., Gabbott, S., Holwell, D., et al. 2014. A chancelloriid-like metazoan from the early Cambrian Chengjiang Lagerstätte, China. *Scientific Reports* **4**, 7340.

Knell, R.J., and Fortey, R.A. 2005. Trilobite spines and beetle horns—sexual selection in the Palaeozoic? *Biology Letter* **1**, 196–9.

Selden, P., and Read, H. 2008. The oldest land animals: Silurian millipedes from Scotland. *Bulletin of the British Myriapod and Isopod Group* **23**, 36–7.

Vermeij, J. 1995. *A Natural History of Shells*. Princeton Science Library.

Yao, H., Dao, M., Imholt, T., Huang, J., Wheeler, K., Bonilla, A., et al. 2010. Protection mechanisms of the iron-plated armour of a deep-sea hydrothermal vent gastropod. *PNAS* **107**, 987–92.

Yoon, C.K. 1995. Scientist at work: Geerat Vermeij; getting the feel of a long ago arms race. *New York Times*. Available at: http://www.nytimes.com/1995/02/07/science/scientist-at-work-geerat-vermeij-getting-the-feel-of-a-long-ago-arms-race.html?pagewanted=all (accessed Aug. 2017).

Chapter 3

Brunet, M., Guy, F., Pilbeam, D., Mackaye, H.T., Likius, A., Ahounta, D., et al. 2002. A new hominid from the Upper Miocene of Chad, Central Africa. *Nature* **418**, 145–51.

Gahn, F.J., and Baumiller, T.K. 2003. Infestation of Middle Devonian (Givetian) camerate crinoids by platyceratid gastropods and its implications for the nature of their biotic interaction. *Lethaia* **36**, 71–82.

Groenewald, G.H., Welman, J., and MacEachern, J.A. 2001. Vertebrate burrow complexes from the Early Triassic Cynognathus Zone (Driekoppen Formation, Beaufort Group) of the Karroo, Basin, South Africa. *Palaios* **16**, 148–60.

Henshilwood, C., d'Errico, F., Yates, R., Jacobs, Z., Tribolo, C., Duller, G.A., et al. 2002. Emergence of modern human behavior: Middle Stone Age engravings from South Africa. *Science* **295**, 1278–80.

Paton, R.L., Smithson, T.R., and Clack, J.A. 1999. An amniote-like skeleton from the Early Carboniferous of Scotland. *Nature* **398**, 508–13.

Shubin, N. 2009. *Your Inner Fish.* Vintage.

Starrfelt, J., and Liow, L.H. 2016. How many dinosaur species were there? Fossil bias and true richness estimated using a Poisson sampling model. *Philosophical Transactions of the Royal Society* **B371**, 20150219.

Villmoare, B., Kimbel, W.H., Seyoum, C., Campisano, C.J., DiMaggio, E.N., Rowan, J., et al. 2015. Early *Homo* at Ledi-Geraru, Afar, Ethiopia. *Science* **347**, 1352–5.

Ward, C.V., Tochen, M.W., Plavcan, J.M., Brown, F.H., and Manthi, F.Y. 2014. Early Pleistocene third metacarpal from Kenya and the evolution of modern human hand-like morphology. *PNAS* **111**, 121–4.

Zalasiewicz, J., and Williams, M. 2014. *Ocean Worlds: The Story of Seas on Earth and Other Planets.* Oxford University Press.

Zamora, S., Rahman, I.A., and Smith, A.B. 2012. Plated Cambrian bilaterians reveal the earliest stages of echinoderm evolution. *PloS One* **7**, e38296.

Chapter 4

Buffon, Comte de. 2018. *The Epochs of Nature.* (Translated and compiled by Zalasiewicz, J., Milon, A.-S., and Zalasiewicz, M., with an introduction by Zalasiewicz, J., Sorlin, S., Robin, L., and Grinevald, J.) Chicago University Press.

Davies, N.S., and Gibling, M.R. 2010. Cambrian to Devonian influence of alluvial systems: The sedimentological impact of the earliest land plants. *Earth Science Reviews* **98**, 171–200.

Davies, N.S., and Gibling, M.R. 2013. The sedimentary record of Carboniferous rivers: Continuing influence of land plant evolution on alluvial processes and Palaeozoic ecosystems. *Earth Science Reviews* **120**, 40–79.

Knauth, L.P., and Kennedy, M.J. 2009. The late Precambrian greening of Earth. *Nature* **460**, 728–32.

Lateiner, D. 2002. Pouring bloody drops (Iliad 16.459): The grief of Zeus. *Colby Quarterly* **38**, 42–61.

Stein, W.E., Mannolini, F., Hernick, L.A., Landing, E., and Berry, C.N. 2007. Giant cladoxylopsid tress resolve the enigma of the Earth's earliest fossil forest at Gilboa. *Nature* **446**, 904–7.

Williams, C.G. 2010. Long-distance pine pollen still germinates after meso-scale dispersal. *American Journal of Botany* **97**, 846–55.

Williams, M., Zalasiewicz, J., Davies, M., Mazzini, I., Goiran, J-P., and Kane, S. 2014. Humans as the third evolutionary stage of biosphere engineering of rivers. *Anthropocene* **7**, 57–63.

Winchester, S. 2001. *The Map that Changed the World: William Smith and the Birth of Modern Geology.* HarperCollins.

Chapter 5

Pálfy, J., and Kocsis, A.T. 2014. Volcanism of the Central Atlantic magmatic province as the trigger of environmental and biotic changes around the Triassic–Jurassic boundary. *Geological Society of America Special Papers* **505**, 245–61.

Chapter 6

Javaux, E.J., Marshall, C.P., and Bekker, A. 2010. Organic-walled microfossils in 3.2-billion-year-old shallow-marine siliciclastic deposits. *Nature* **463**, 934–8.

Zalasiewicz, J., and Williams, M. 2014. *Ocean Worlds: The Story of Seas on Earth and Other Planets.* Oxford University Press.

Chapter 7

Balter, M. 2015. Feathered fossils from China reveal dawn of modern birds. *Science News.* Available at: http://www.sciencemag.org/news/2015/05/feathered-fossils-china-reveal-dawn-modern-birds (accessed Aug. 2017).

Bressan, D. 2014. Bat-pterodactyls. *Scientific American.* Available at: https://blogs.scientificamerican.com/history-of-geology/bat-pterodactyls (accessed Aug. 2017).

Chapelle, G., and Peck, L.S. 1999. Polar gigantism dictated by oxygen availability. *Nature* **399**, 114–15.

Engel, M.S., and Grimaldi, D.A. 2004. New light shed on the oldest insect. *Nature* **427**, 627–30.

Hone, D. 2009. What on Earth are pycnofibers? *Archosaur Musings.* Available at: https://archosaurmusings.wordpress.com/2009/08/05/what-on-earth-are-pycnofibers/ (accessed Aug. 2017).

Hu, G., Lim, K.S., Horvitz, N., Clark, S.J., Reynolds, D.R., Sapir, N., et al. 2016. Mass seasonal bioflows of high-flying insect migrants. *Science* **354**, 1584–7.

McCracken, G.F., Safi, K., Kunz, T.H., Dechmann, D.K., Swartz, S.M., and Wikelski, M. 2016. Airplane tracking documents the fastest flight speeds recorded for bats. *Royal Society Open Science* **3**, 160398.

Meng, J., Hu, Y., Wang, Y., Wang, X., and Li, C. 2006. A Mesozoic gliding mammal from northeastern China. *Nature* **444**, 889–93.

Morris, H. 2017. How many planes are there in the world right now? *Telegraph*. Available at: http://www.telegraph.co.uk/travel/travel-truths/how-many-planes-are-there-in-the-world/ (accessed Aug. 2017).

Norberg, U.M.L. 2006. Evolution of flight in animals. *WIT Transactions on State of the Art in Science and Engineering*, vol. 3. WIT Press.

Trewin, N.H. 2004. History of research on the geology and palaeontology of the Rhynie area, Aberdeenshire, Scotland. *Transactions of the Royal Society of Edinburgh: Earth Sciences* **94**, 285–97.

Witton, M.P. 2015. Were early pterosaurs inept terrestrial locomotors? *PeerJ* **3**, e1018.

Xu, G.H., Zhao, L.J., Gao, K.Q., and Wu, F.X. 2013. A new stem-neopterygian fish from the Middle Triassic of China shows the earliest over-water gliding strategy of the vertebrates. *Proceedings Biological Sciences* **280**, 20122261.

Chapter 8

Carson, R. 2000. *Silent Spring*, reprinted. Penguin.

Du Hamel and de Buffon. 1737. De la cause de l'excentricité des couches ligneuses qu'on apperçoit quand on coupe horisontalement le tronc d'un arbre; de l'inégalité d'épaisseur, & de different nombre de ces couches, tant dans le bois formé que dans l'aubier. *Histoire de l'Académie Royale des Sciences*, 121–34.

Hays, J.D., Imbrie, J., and Shackleton, N.J. 1976. Variations in the Earth's orbit: pacemaker of the ice ages. *Science* **194**, 1121–32.

Lamb, A.L., Evans, J.A., Buckley, R., and Appleby, J. 2014. Multi-isotope analysis demonstrates significant lifestyle changes in King Richard III. *Journal of Archaeological Science* **50**, 539–65.

Mauritian Wildlife Foundation. 2017. Tortoise re-wilding. Available at: http://www.mauritian-wildlife.org/application/index.php?tpid=30&tcid=80 (accessed Aug. 2017).

Morelle, R. 2013. Clam-gate: the Epic Saga of Ming. Available at: http://www.bbc.co.uk/news/science-environment-24946983 (accessed Aug. 2017).

Schopf, J.M. 1970. Petrified peat from a Permian coal bed in Antarctica. *Science,* **169**, 274–7.

Scrutton, C.T. 1964. Periodicity in coral growth. *Palaeontology* 7, 552–8.

Tey, J. 2009. *The Daughter of Time*, reprint. Arrow.

Trotter, J.A., Williams, I.S., Barnes, C.R., Lécuyer, C., and Nicoll, R.S. 2008. Did cooling oceans trigger Ordovician biodiversification? Evidence from conodont thermometry. *Science* **321**, 550–4.

Wells, J. 1963. Coral growth and geochronometry. *Nature* **197**, 948–50.

Chapter 9

Barnosky, A.D., Matzke, N., Tomiya, S., Wogan, G.O., Swartz, B., Quental, T.B., et al. 2011. Has the Earth's sixth mass extinction already arrived? *Nature* **471**, 51–7.

Dixon, D. 1983. *After Man*. St Martin's Press.

Koch, P.L., and Barnosky, A.D. 2006. Late Quaternary extinctions: State of the debate. *Annual Review of Ecology, Evolution and Systematics* **37**, 215–50.

Mayor, A. 2000. *The First Fossil Hunters: Palaeontology in Greek and Roman Times*. Princeton University Press.

Ross, J. 2017. *Darwin's Pigeons*. Available at: http://darwinspigeons.com/ (accessed Aug. 2017).

Smil, V. 2011. Harvesting the biosphere. *Population and Development Review* **37**, 613–36.

Chapter 10

Adelmann, J. 2007. *Eozoön:* Debunking the dawn animal. *Endeavour* **31**, 94–8.

Allwood, A.C., Walter, M.R., Kamber, B.S., Marshall, C.P., and Burch, I.W. 2006. Stromatolite reef from the Early Archaean era of Australia. *Nature* **441**, 714–18.

Gehler, A., and Reich, M. 2008. Ernst Louis Kalkowsky (1851–1938) and the term 'stromatolite'. In: Reitner, J., Quéric, N-V., and Reich, M. (eds). *Geobiology of Stromatolites*. University of Gottingen, 9–17.

Logan, B.W. 1961. Cryptozoon and associated stromatolites from the recent, Shark Bay, Western Australia. *Journal of Geology* **69**, 517–33.

O'Brien, C.F. 1970. *Eozoön canadense* 'The Dawn Animal of Canada'. *Isis* **61**, 206–23.

Riding, R. 2011. The nature of stromatolites: 3500 million years of history and a century of research. In: Reitner, J., Quéric, N.V., Arp, G. (eds) *Advances in Stromatolite Biology, Lecture Notes in Earth Sciences*, Vol. **131**. Springer, 29–74.

Wacey, D., Kilburn, M.R., Saunders, M., Cliff, J., and Brasier, M. 2011. Microfossils of sulphur-metabolizing cells in 3.4 billion-year-old rocks of Western Australia. *Nature Geoscience* **4**, 698–702.

Wadsworth, J., and Cockell, C.S. 2017. Perchlorates on Mars enhance the bactericidal properties of UV light. *Scientific Reports* **7**, 4662.

Zalasiewicz, J., and Williams, M. 2014. *Ocean Worlds: The Story of Seas on Earth and Other Planets*. Oxford University Press.

FIGURE CREDITS

1. Iván Cortijo.
2. Adrian Rushton, John Ahlgren.
3. Derek Siveter, Xianguang Hou.
4. Derek Siveter, Xianguang Hou, Peiyun Cong, Xiaoya Ma.
5. Derek Siveter, Xianguang Hou.
6. Derek Siveter, Xianguang Hou.
7. Photo by Rob Stothard/Getty Images.
8. Tom Harvey.
9. Derek Siveter, Xianguang Hou.
10. Derek Siveter.
11. Jason Dunlop.
12. Jan Zalasiewicz.
13. Daderot/Wikimedia Commons/Public Domain.
14. Derek Siveter, Xianguang Hou.
15. Mark Purnell.
16. Mark Purnell.
17. Reconstruction by Bashford Dean in 1909. Public Domain.
18. Figure 4 of Zhu M, Yu X, Choo B, Qu Q, Jia L, et al. (2012) Fossil fishes from China provide first evidence of dermal pelvic girdles in osteichthyans. *PLoS ONE* 7(4): e35103. Brian Choo/CC-BY-SA-2.5.
19. Nobu Tamura/CC-BY-SA-4.0.
20. DiBgd at English Wikipedia/CC-BY-SA-3.0.
21. National Museum Cardiff/Wikimedia Commons/Public Domain.
22. Didier Descouens/CC-BY-SA-4.0.
23. Scott Cuper/CC-BY-SA-4.0.
24. Ulrich Salzmann.
25. Ulrich Salzmann.
26. NASA/World History Archive/AGE Fotostock.
27. NASA Earth Observatory image created by Jesse Allen and Robert Simmon, using EO-1 ALI data provided courtesy of the NASA EO-1 team.
28. Chris Nedza.
29. Ian Wilkinson.
30. Vincent Perrier.
31. David Siveter.
32. Jeremy Young.

33. © Paulo de Oliveira/NHPA/AGE Fotostock.
34. (A) Description by Italian naturalist Cosimo A. Collini:
    *Sur quelques zoolithes du cabinet d´histoire naturelle de
    S.A.S.E. palatine et de Baviere, à Mannheim.* In: Acta Academiae Theodoro-Palatinae,
    vol. 5 Phys., Mannheim 1784, pp. 58–71. Engraving by Egid Verhelst/Public
    Domain. (B) David Martill.
35. © Ivan Kuzmin/Shutterstock.com.
36. Bernd Schöne and Soraya Marali.
37. Bjørn Christian Tørrissen/ CC-BY-SA-4.0.
38. Chris Nedza.
39. The Library of Congress/Public Domain.
40. © University of Leicester.
41. Chroniques d'Enguerrand de Monstrelet (early 15th century)/Public
    Domain.
42. NASA.
43. Happy Little Nomad/CC-BY-SA-2.0.

INDEX

pterosaur xi, 88, 181, 182, 184–6, 188, 189, 191, 192
pulp cavity 63
Punch 95
pycnofibres 184, 185, 189
pyramid, Egyptian 147
pyrite 260
pyroxene 249
python 231

quadrate bone 93
quahog, ocean 197, 198, 200, 201
Quarterly Journal of the Geological Society of London 250
quartz 164
Queen Alexandra Range 220
Queen Elizabeth I 149, 240
quelea, red-billed 237
Quetzalcoatlus 184

rabbit 102
radiolarian 165–8
radiometric ages 9
radula 48, 49
rat 226, 227
rat, brown 237
raven 189
ray-fin fish 76, 77, 78
Réamur, René-Antoine Ferchault de 130
redwood tree 108
Red Jungle Fowl 236
reef gap 142
reef, non-coral-dominated 145–7
rejectamenta 44
Reophax 155
Repenomamus 226
reptile 51, 61, 69, 76, 81, 82, 97, 180, 181, 190
reptile, marine xii, 87, 91, 96
Rhamphorhynchus 185
rhinoceros, woolly 229
rhizopod 166
Rhynie Chert 178

Rhyniognatha hirsti 178, 180
Rhysonetron 250
rice 126
Richard III 222–4
Richard the Lionheart 124
Rickards, Barrie 57
Ripley 246
river, meandering 117, 118
river, braided 117, 118
robot 243
rock dove 235
Roman Empire 236
Roman peoples 11, 97, 227, 240
Romania 184
Romer, Alfred 80
Romer's Gap 80
rotalinid 155
Rowe, Arthur 105
Rowney, Thomas 250
Royal Navy 157
Royal Society 130, 150–2, 156, 158
Royal Society of British Sculptors 158
rudist bivalve 143, 144, 233
rudist reef 143, 144
Rugen Island 170
Ruhmkorff apparatus 129
'Russian thistle' 121
Rusophycus 34

sabre-tooth tiger 229
Sahelanthropus 98, 99
St Bernard dog 235
St Paul's Cathedral 151
Salter, John William 218, 219
Sand 18
Santiago de Cuba 240
sapwood 110
Sarpedon 124
saurischian 86
sauropod 86, 87, 227
sauropsid 82, 85
Saxony 253
Saxony, Lower 253

worm 6, 15, 16, 31, 32, 45, 130, 208
worm, polychaete 26
worm, priapulid 7, 16, 26, 59
worm tooth 68
worm, velvet 26, 27, 32
Wortmania 226
Wrangel Island 12
wren 189
Wren, Sir Christopher 51, 151
Wright, Orville 195
Wrisberg, Heinrich August 152
Wyndham, John 162
Wyoming 192

X-ray 246

Yarra River 112
'year without a summer' 199, 234
Yorkshire 51
Yucatan 225
Yukon river 117
Yunnan Province 13, 15, 16, 17, 30, 66

Zeus 124
Zhou, Min 76
Zinc 146
zircon 9
Zoological Museum, Kiel 198
zoophyte 129
zooplankton 166
zooxanthellae 132, 231